Grasses of the Southwest

A KEY TO COMMON SPECIES USING VEGETATIVE FEATURES

BLUE GRAMA
Bouteloua gracilis

STEVE W. CHADDE

GRASSES OF THE SOUTHWEST
A Key to Common Species Using Vegetative Features

STEVE W. CHADDE

The Biota of North America Program (*www.bonap.org*) provided permission to use their data to generate the distribution maps.

The author can be reached at: *steve@chadde.net*

AN ORCHARD INNOVATIONS BOOK
ISBN: 978-1951682347

Ver. 1.0 (04/17/2020)

CONTENTS

CONTENTS

INTRODUCTION

Native and cultivated grasses must often be identified when plants are immature or have been heavily grazed and, thus, do not have flower stalks. Under such conditions, conventional botanical manuals offer little assistance. Examination of vegetative characteristics, however, will often permit tentative identification until better developed specimens are available.

The purpose of this paper is to present an illustrated key for the identification of 73 common range and pasture grasses of the southwestern Unied States by vegetative characters.

A technical knowledge of taxonomic botany, although useful, is not essential to use this method of identification. A magnifying glass or hand lens of 5 to 10 power, plus a small millimeter scale (also provided on page 179), will prove helpful.

Two recent publications can be consulted for more information on these and other grasses of the southwestern region: *Manual of Grasses for North America* (2015), and *Grasses of the Intermountain Region* (2015). Keys, maps, and descriptions of North American grass species are also available online at: *http://herbarium.usu.edu/webmanual*. Scientific names used in this guide follow those of the above mentioned *Manual of Grasses for North America.*

Portions of this guide were originally published as *A Vegetative Key to Some Common Arizona Range Grasses* (General Technical, Report RM-53, February 1978) by R. F. Copple and C. P. Pase, of the Rocky Mountain Forest and Range Experiment Station, USDA Forest Service.

DESCRIPTIONS OF VEGETATIVE CHARACTERS

Annual or perennial – Annual grasses generally have fine, poorly developed root systems, and no evidence of old, previous year's culm bases. Perennial grass root systems are better developed, and evidence of last year's leaves or flower stalks can usually be found. Presence of creeping root stalks, rhizomes, stolons, etc., always indicates a perennial grass. Seedling perennial grasses may occasionally be confused with annuals. Some caution should be used with this character in short-lived perennials such as *Bouteloua barbata* (fig. 1).

Vernation – This refers to the cross section of a grass blade in a developing shoot. Vernation may be best determined by cutting across the sheath just below the collar, and observing the shape of the enclosed cut blade with a strong hand lens. Care should be used to make the cross section with a razor blade or sharp knife, to avoid mashing the enclosed blade and destroying its original shape. Vernation may be folded, rolled (or curled), or clasped (fig. 2).

Folded vernation is generally associated with a flattened sheath, and rolled or clasped with a rounded sheath, but this does not always hold true.

Blades – Width and length of blades vary somewhat, depending on growing conditions, but within limits offer an easy method to separate certain species. Shaded plants have longer and wider blades than those exposed to the sun.

INTRODUCTION

Nerves may be raised on either surface—sometimes prominently, as on the upper surface of *Koeleria macrantha* blades.

Blades may be folded, flattened, or rounded. Normally flat blades may tend to appear round as they dry, but examination of the base of the blade above the sheath will reveal its normal condition. Blade tips may be boat-shaped, as in some species of *Poa*.

Grass blades are two-ranked, while sedges are three-ranked (fig. 1).

Auricle – Small, membranous extensions of the collar of some grasses, notably of *Pascopyron*, *Lolium*, and others of the tribe Hordeae (fig. 2).

Ligule – This is an organ clasping the stem at the junction of the blade and sheath. It may be membranous or hairy, or both, as in truncate-ciliate ligules. Size, shape, and type of ligule is quite constant, and provides one of the best vegetative characters for identification (fig. 3).

Collar – This is the junction of leaf blade and sheath. It may be a broad or narrow band, glabrous or covered with fine hair. The margin may be glabrous, hairy, or with glandular hairs.

Sheath – The tubular basal portion of the leaf, below the blade that clasps the stem. The sheath may be closed, as in *Bromus*, or open. Ruptured closed sheaths should not be confused with normally open ones (fig. 2).

The sheath of certain grasses may have a thin, membranaceous, hyaline margin.

Internodes – The area between successive nodes.

Nodes – The joints of a grass stem or culm.

Roots – Grass root systems are fibrous, but rhizomes or thickened underground stems may also be present.

Culm – The flowering stem of a grass, generally round or elliptical in cross section (fig. 1) as opposed to the triangular stems of sedges. Rushes have pithy stems, while most grasses have hollow stems.

Considerable variation may be expected in certain characters such as degree of hairiness of the blade or sheath, length and width of blade, etc. It may therefore be necessary to examine several specimens before deciding what is a representative condition. Other characteristics, such as type of ligule and vernation, are generally rather constant.

GLOSSARY

ACUTE: Terminating sharply in an angle of less than 90 degrees.

AURICLE: Small, membranous, claw-like extension of the collar of some grasses.

AWN: A needle-like extension of the grass floret, as in *Hesperostipa* (needlegrass) or *Aristida* (three-awn).

CANESCENT: Grayish or whitish due to dense, fine hairs.

CILIATE: Fringed with hairs on the edge.

COLLAR: The region on the outer side of the grass leaf at the junction of the sheath and blade.

CORM: A swollen or enlarged, rounded, solid, fleshy, mostly subterranean, stem base. Like a bulb in shape and appearance, except that it is solid instead of being composed of fleshy scales.

CULM: The jointed stalk or stem of grasses, usually hollow except at the nodes, and mostly herbaceous.

DECUMBENT: Reclining on the ground but with the end ascending, bending horizontally at the base. Said of stems. Decumbent conveys the idea of weakness.

DIVIDED: Refers to the shape of the collar on the back of the sheath.

DORSAL: Refers to the top of the blade, opposite to ventral.

ELLIPTICAL: The cross section of the stem partially flattened.

ENTIRE: With an even margin, without teeth or lobes.

FIBROUS: Refers to a mass of thread-like roots.

FLORET: A unit of the spikelet of a grass inflorescence, consisting usually of lemma, palea, stamens, and pistil.

GENICULATE: Knee-like; bent like a knee; said of stems and awns of needlegrass (*Hesperostipa*).

GLABRATE: Becoming glabrous or nearly so in age.

GLABROUS: Devoid of hairs or pubescence; smooth in the sense of absence of all hairiness.

GLANDULAR: Pertaining to or possessing glands at the base of hairs.

GLOSSY: Smooth, shining.

HIRSUTE: Hairy with rather coarse, stiffish, straight, beardlike hairs.

HISPID: Bristly; beset with stiff, rough, bristle-like hairs.

HYALINE: Thin and translucent.

INFLORESCENCE: The flowering portion of a grass plant, consisting of spikelets and florets.

INTERNODE: The portions of a stem between the nodes, or joints.

INVOLUTE: Inrolled; i.e. with both edges rolled in toward the middle (blade) each edge presenting a spiral appearance in cross-section.

LACERATE: Deeply and irregularly cut along the edges.

LANCEOLATE: Lance-shaped; several times longer than broad, and tapering from the relatively narrow base to apex.

LIGULE: The projecting, usually tongue-like, membranous end of the lining of the leaf sheath, seen at the base of the leaf, between it and the stalk, and a very characteristic feature of the grass family. The ligule is quite often an important means of distinguishing grasses; sometimes it is reduced to a mere fringe of hairs or to a hardened ring.

GLOSSARY

MAT: Growing thickly, or closely interwoven.

MIDRIB: The central or main rib common in blades of grasses.

NODE: A joint or knot. Said especially of stems, whose nodes or joints are enlarged, often dark colored, and are the points whence leaves and additional flowers often spring.

NOTCHED: An edge with teeth or lobes.

OBTUSE: Blunt or rounded at the tip; not sharply pointed.

PANICLE: An open inflorescence in which the lower branches are typically longer and blossom earlier than the upper branches.

PERENNIAL: Lasting for three or more years; said especially of herbaceous plants that are neither annual nor biennial.

PILOSE: Hairy with soft, slender hairs.

PLUMED: Feather-like, having fine hairs on either side as the awn of certain *Hesperotipa* species.

PUBESCENT: Clothed with soft hairs or down.

PUNGENT: Sharply and rigidly pointed.

RACEME: A simple, elongated, indeterminate flower cluster.

RACHIS: The axis of a spike, raceme, or branch of a panicle; the organ that supports the florets.

RETRORSE: Directed back or downward, as hairs on the sheath of certain grasses.

RHIZOME: An underground stem; recognized by the regular nodes and internodes; e.g. *Sorghum halepense.*

RIB: The primary veins of a blade.

RUDIMENTARY: Reduced.

RUNNER: A long, slender form of creeping branch, prostrate on the ground. Each runner, after having grown to its full length, strikes root from the tip (it sometimes roots at the joints also, in which case it may merge into a stolon), fixing the tip to the ground, then forms a bud at that point, which later develops into a tuft of leaves and so gives rise to a new plant.

SCABROUS: Rough or harsh to the touch.

SCALLOPED: Margin of a blade that buckles, resulting in waves.

SHEATH: In grasses, the lower part of the leaf which envelops the stem or culm.

SHREDDED: Deeply and irregularly cut along the edges.

SOD: A group of plants growing densely together forming a solid mat, as buffalograss or Kentucky bluegrass grown for lawns.

SOLID: Applied to the culm, which is solid in a few grasses and all sedges and rushes.

SPIKELET: In grasses, the name applied to the cluster of one or more florets.

STOLON: A trailing or reclining branch, above ground, which takes root where it touches the soil, there sending up new shoots which later become separate plants.

TOOTHED: An edge with regular notches.

TRUNCATE: Squared at the tip; terminating abruptly as if cut off crosswise.

GLOSSARY

VEIN: One of the fibrovascular bundles forming part of the framework (skeleton) of a leaf; more or less parallel veinations (as in a grass blade).

VENTRAL: Refers to the inner surface of the blade, facing the culm as the leaf is held erect.

VILLOUS: Densely hairy with long, soft hairs.

WOOLLY: Covered with long, matted hairs.

rhizome

crown

FIBROUS

RHIZOMATOUS

stolon

stolon internode

scale leaf

STOLONIFEROUS

GRASS IDENTIFICATION CHARACTERS

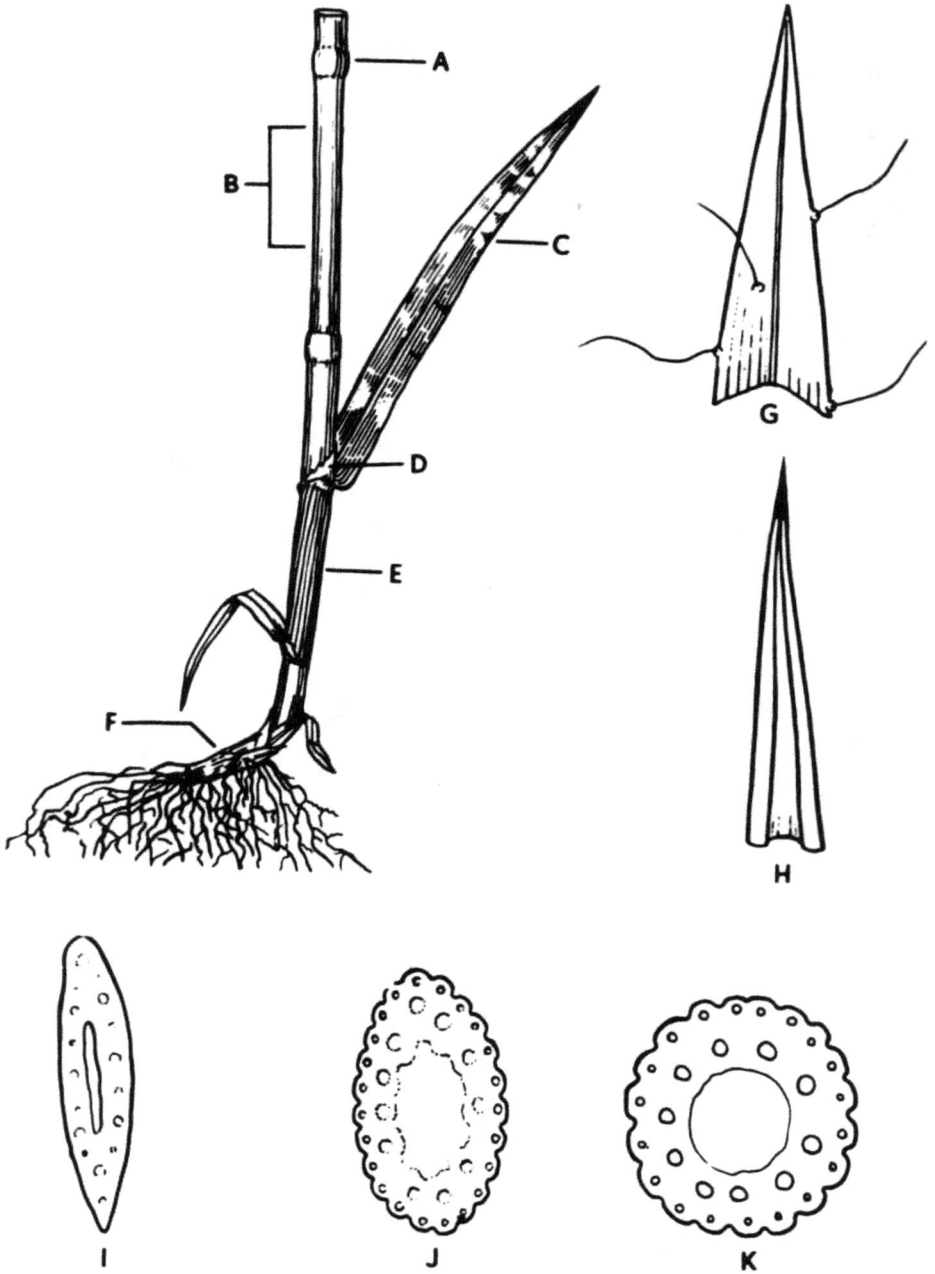

FIGURE 1. *Parts of typical grass plant:* A, node; B, internode; C, blade; D, ligule; E, sheath; F, rhizome. *Leaf blades:* G, with glandular hairs; H, with pungent tip. *Culm shape in cross section:* I, flat; J ,elliptical; K, round.

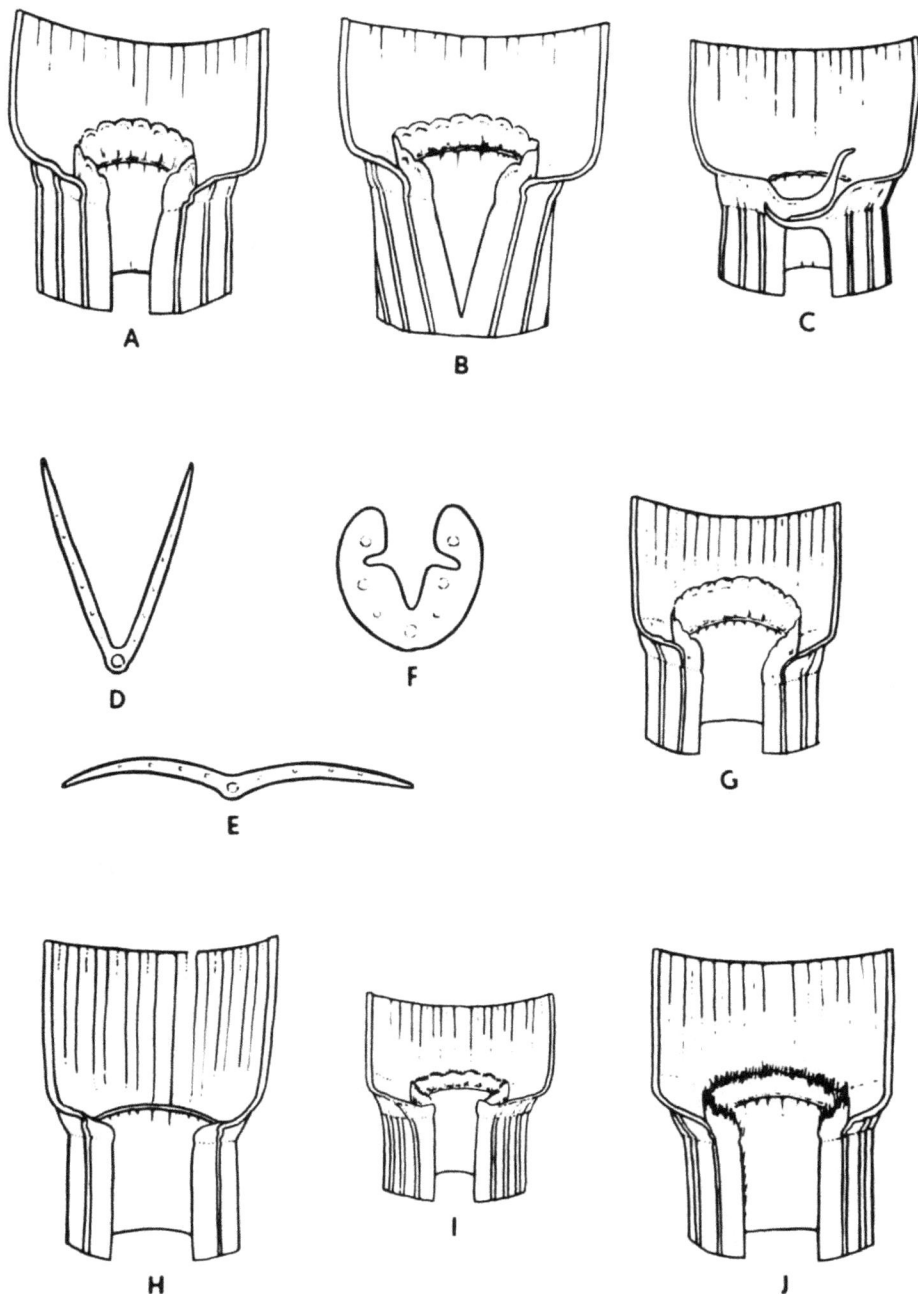

FIGURE 2. *Sheaths:* A, open; B, closed; C, with clasping auricle. *Leaf shape in cross section:* D, folded; E, flat; F, round. *Ligule types:* G, membranous; H, collar-like; I, hairy; J, truncate-ciliate (membranous with ciliate margin).

GRASS IDENTIFICATION CHARACTERS

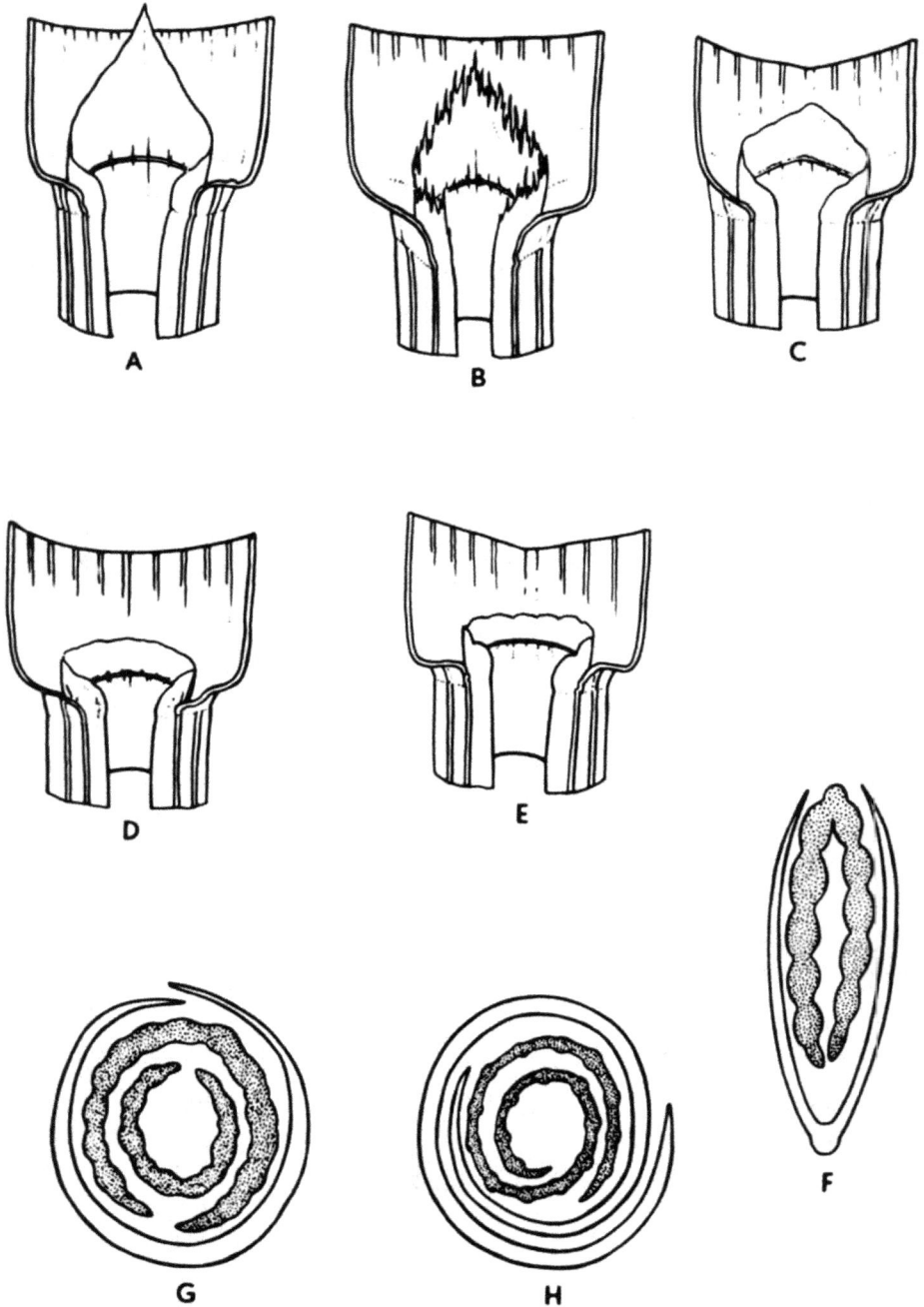

FIGURE 3. *Membranous ligule shapes:* A, acute; B, acute-lacerate; C, obtuse; D, truncate; E, truncate-notched. *Vernation* (cross section of leaf in bud): F, folded; G, clasped; H, curled or rolled.

KEY CHARACTERS

1. **AURICLES**
   2. *Agropyron cristatum*
      CRESTED WHEATGRASS
   31. *Elymus elymoides*
      BOTTLEBRUSH SQUIRRELTAIL
   44. *Lolium perenne*
      PERENNIAL RYEGRASS
   55. *Pascopyrum smithii*
      WESTERN WHEATGRASS

2. **STOLONS**
   26. *Cynodon dactylon*
      BERMUDAGRASS
   29. *Distichlis spicata*
      DESERT SALTGRASS
   37. *Hilaria belangeri*
      CURLY-MESQUITE
   41. *Hopia obtusa*
      VINE-MESQUITE

3. **NODES HAIRY OR PUBESCENT**
   10. *Bothriochloa barbinodis*
      CANE BLUESTEM
   28. *Digitaria californica*
      ARIZONA COTTONTOP
   37. *Hilaria belangeri*
      CURLY-MESQUITE
   40. *Hilaria rigida*
      BIG GALLETA
   50. *Muhlenbergia pungens*
      SANDHILL MUHLY
   65. *Setaria parviflora*
      KNOTROOT BRISTLEGRASS
   72. *Tridens muticus*
      SLIM TRIDENS

4. **BLADE TIP VERY SHARP**
   50. *Muhlenbergia pungens*
      SANDHILL MUHLY

5. **FLOWERS CLUSTERED AT APEX**
   27. *Dasyochloa pulchella*
      FLUFFGRASS

6. **CORMS OR BULBS AT BASE OF CULMS**
   73. *Zuloagaea bulbosa*
      BULB PANICUM

7. **SHEATH MARGIN WHITE**
   45. *Lycurus phleoides*
      WOLFTAIL
   51. *Muhlenbergia richardsonis*
      MAT MUHLY
   60. *Schedonnardus paniculatus*
      TUMBLEGRASS

8. **STEMS OR CULMS ROOTING AT NODES**
   15. *Bouteloua eriopoda*
      BLACK GRAMA
   27. *Dasyochloa pulchella*
      FLUFFGRASS
   37. *Hilaria belangeri*
      CURLY-MESQUITE

9. **RACHIS EXTENDS BEYOND SPIKELETS**
   17. *Bouteloua hirsuta*
      HAIRY GRAMA

10. **SHEATH CLOSED, PARTIALLY OR COMPLETELY**
    19. *Bromus anomalus*
      NODDING BROME
    20. *Bromus arizonicus*
      ARIZONA BROME
    21. *Bromus ciliatus*
      FRINGED BROME
    22. *Bromus inermis*
      SMOOTH BROME
    23. *Bromus rubens*
      RED BROME
    24. *Bromus tectorum*
      DOWNY CHESS,
      CHEATCRASS BROME

KEY CHARACTERS

11. FLOWERS INSERTED IN SHEATH
68. *Sporobolus compositus*
    TALL DROPSEED
69. *Sporobolus cryptandrus*
    SAND DROPSEED

12. INTERNODE HAIRY
15. *Bouteloua eriopoda*
    BLACK GRAMA
30. *Elionurus barbiculmis*
    WOOLSPIKE
40. *Hilaria rigida*
    BIG GALLETA
65. *Setaria parviflora*
    KNOTROOT BRISTLEGRASS

13. BLADE MARGIN WITH GLANDULAR HAIRS
11. *Bouteloua aristidoides*
    NEEDLE GRAMA
13. *Bouteloua chondrosioides*
    SPRUCETOP GRAMA
14. *Bouteloua curtipendula*
    SIDEOATS GRAMA
15. *Bouteloua eriopoda*
    BLACK GRAMA
17. *Bouteloua hirsuta*
    HAIRY GRAMA
18. *Bouteloua repens*
    SLENDER GRAMA
36. *Heteropogon contortus*
    TANGLEHEAD
37. *Hilaria belangeri*
    CURLY-MESQUITE
73. *Zuloagaea bulbosa*
    BULB PANICUM

14. GROWTH A MAT OR SOD-LIKE
16. *Bouteloua gracilis*
    BLUE GRAMA
26. *Cynodon dactylon*
    BERMUDAGRASS

37. *Hilaria belangeri*
    CURLY-MESQUITE
51. *Muhlenbergia richardsonis*
    MAT MUHLY
53. *Muhlenbergia torreyi*
    RING MUHLY
55. *Pascopyrum smithii*
    WESTERN WHEATGRASS
73. *Zuloagaea bulbosa*
    BULB PANICUM

15. BLADE MARGIN WHITE
10. *Bothriochloa barbinodis*
    CANE BLUESTEM
28. *Digitaria californica*
    ARIZONA COTTONTOP
45. *Lycurus phleoides*
    WOLFTAIL
66. *Sorghum halepense*
    JOHNSONGRASS

16. BLADE FOLDED
44. *Lolium perenne*
    PERENNIAL RYEGRASS
45. *Lycurus phleoides*
    WOLFTAIL
47. *Muhlenbergia emersleyi*
    BULLGRASS
53. *Muhlenbergia torreyi*
    RING MUHLY
58. *Poa fendleriana*
    MUTTON BLUEGRASS
60. *Schedonnardus paniculatus*
    TUMBLEGRASS
62. *Schizachyrium cirratum*
    TEXAS BLUESTEM
63. *Schizachyrium scoparium*
    LITTLE BLUESTEM
65. *Setaria parviflora*
    KNOTROOT BRISTLEGRASS
73. *Zuloagaea bulbosa*
    BULB PANICUM

KEY CHARACTERS

17. RHIZOMES
3. *Agrostis gigantea*
   REDTOP
22. *Bromus inermis*
   SMOOTH BROME
26. *Cynodon dactylon*
   BERMUDAGRASS
29. *Distichlis spicata*
   DESERT SALTGRASS
38. *Hilaria jamesii*
   GALLETA
39. *Hilaria mutica*
   TOBOSA
51. *Muhlenbergia richardsonis*
   MAT MUHLY
55. *Pascopyrum smithii*
   WESTERN WHEATGRASS
59. *Poa pratensis*
   KENTUCKY BLUEGRASS
66. *Sorghum halepense*
   JOHNSONGRASS

18. BLADE GLOSSY VENTRALLY
44. *Lolium perenne*
   PERENNIAL RYEGRASS

**19. SHEATH BASE COLOR
USUALLY PINK**
44. *Lolium perenne*
   PERENNIAL RYEGRASS

**20. BLADE A BRISTLE,
NEEDLE-LIKE, ROUND**
1. *Achnatherum hymenoides*
   INDIAN RICEGRASS
5. *Aristida californica*
   SANTA RITA THREE-AWN
6. *Aristida divaricata*
   POVERTY THREE-AWN
7. *Aristida purpurea*
   BLUE THREE-AWN
8. *Aristida ternipes*
   SPIDERGRASS
9. *Blepharoneuron tricholepis*
   PINE DROPSEED

27. *Dasyochloa pulchella*
   FLUFFGRASS
30. *Elionurus barbiculmis*
   WOOLSPIKE
34. *Festuca arizonica*
   ARIZONA FESCUE
35. *Hesperostipa neomexicana*
   NEW MEXICAN FEATHERGRASS
46. *Muhlenbergia curtifolia*
   UTAH MUHLY
48. *Muhlenbergia montana*
   MOUNTAIN MUHLY
50. *Muhlenbergia pungens*
   SANDHILL MUHLY
51. *Muhlenbergia richardsonis*
   MAT MUHLY
52. *Muhlenbergia rigens*
   DEERGRASS
61. *Schismus barbatus*
   MEDITERRANEANGRASS
67. *Sporobolus airoides*
   ALKALI SACATON
72. *Tridens muticus*
   SLIM TRIDENS

21. LETTER M OR W ON BLADE
22. *Bromus inermis*
   SMOOTH BROME

**22. SHEATH USUALLY WITH
GLANDULAR HAIRS**
41. *Hopia obtusa*
   VINE-MESQUITE
73. *Zuloagaea bulbosa*
   BULB PANICUM

23. SEEDS THREE-AWNED
ONE LONG AND TWO SHORT AWNS
8. *Aristida ternipes*
   SPIDERGRASS

THREE AWNS OF EQUAL LENGTH
4. *Aristida adscensionis*
   SIXWEEKS THREE-AWN
5. *Aristida californica*
   SANTA RITA THREE-AWN

KEY CHARACTERS

23. SEEDS THREE-AWNED *cont.*

6. *Aristida divaricata*
   POVERTY THREE-AWN
7. *Aristida purpurea*
   BLUE THREE-AWN

24. SINGLE AWN

A. AWN SHORT (1 INCH OR LESS)

10. *Bothriochloa barbinodis*
    CANE BLUESTEM
24. *Bromus tectorum*
    DOWNY CHESS,
    CHEATCRASS BROME
25. *Chloris virgata*
    FEATHER FINGERGRASS
47. *Muhlenbergia emersleyi*
    BULLGRASS
48. *Muhlenbergia montana*
    MOUNTAIN MUHLY
49. *Muhlenbergia porteri*
    BUSH MUHLY
50. *Muhlenbergia pungens*
    SANDHILL MUHLY
66. *Sorghum halepense*
    JOHNSONGRASS

B. AWN LONG (MORE THAN 1 IN.)

AWN STRAIGHT

31. *Elymus elymoides*
    BOTTLEBRUSH SQUIRRELTAIL

AWN TWISTED OR GENICULATE

35. *Hesperostipa neomexicana*
    NEW MEXICAN FEATHERGRASS
36. *Heteropogon contortus*
    TANGLEHEAD
56. *Piptochaetium pringlei*
    PRINGLE NEEDLEGRASS

25. STEMS FLAT

10. *Bothriochloa barbinodis*
    CANE BLUESTEM
32. *Eragrostis intermedia*
    PLAINS LOVEGRASS
36. *Heteropogon contortus*
    TANGLEHEAD

43. *Leptochloa dubia*
    GREEN SPRANGLETOP
45. *Lycurus phleoides*
    WOLFTAIL
53. *Muhlenbergia torreyi*
    RING MUHLY
58. *Poa fendleriana*
    MUTTON BLUEGRASS
59. *Poa pratensis*
    KENTUCKY BLUEGRASS
60. *Schedonnardus paniculatus*
    TUMBLEGRASS
63. *Schizachyrium scoparium*
    LITTLE BLUESTEM
64. *Setaria macrostachya*
    PLAINS BRISTLEGRASS
73. *Zuloagaea bulbosa*
    BULB PANICUM

26. CULMS USUALLY BRANCHED

3. *Agrostis gigantea*
   REDTOP
4. *Aristida adscensionis*
   SIXWEEKS THREE-AWN
5. *Aristida californica*
   SANTA RITA THREE-AWN
7. *Aristida purpurea*
   BLUE THREE-AWN
8. *Aristida ternipes*
   SPIDERGRASS
10. *Bothriochloa barbinodis*
    CANE BLUESTEM
11. *Bouteloua aristidoides*
    NEEDLE GRAMA
14. *Bouteloua curtipendula*
    SIDEOATS GRAMA
25. *Chloris virgata*
    FEATHER FINGERGRASS
26. *Cynodon dactylon*
    BERMUDAGRASS
27. *Dasyochloa pulchella*
    FLUFFGRASS
28. *Digitaria californica*
    ARIZONA COTTONTOP

KEY CHARACTERS

30. *Elionurus barbiculmis*
   WOOLSPIKE
33. *Eragrostis lehmanniana*
   LEHMANN LOVEGRASS
37. *Hilaria belangeri*
   CURLY-MESQUITE
40. *Hilaria rigida*
   BIG GALLETA
45. *Lycurus phleoides*
   WOLFTAIL
49. *Muhlenbergia porteri*
   BUSH MUHLY
54. *Pappophorum vaginatum*
   FEATHER PAPPUSGRASS
65. *Setaria parviflora*
   KNOTROOT BRISTLEGRASS
66. *Sorghum halepense*
   JOHNSONGRASS
68. *Sporobolus compositus*
   TALL DROPSEED
69. *Sporobolus cryptandrus*
   SAND DROPSEED

27. LIGULE LARGE, ACUTE

LIGULE 2–5 MM LONG
20. *Bromus arizonicus*
   ARIZONA BROME
32. *Eragrostis intermedia*
   PLAINS LOVEGRASS
45. *Lycurus phleoides*
   WOLFTAIL
58. *Poa fendleriana*
   MUTTON BLUEGRASS

LIGULE 5–8 MM LONG
48. *Muhlenbergia montana*
   MOUNTAIN MUHLY

LIGULE 8–15 MM LONG
47. *Muhlenbergia emersleyi*
   BULLGRASS

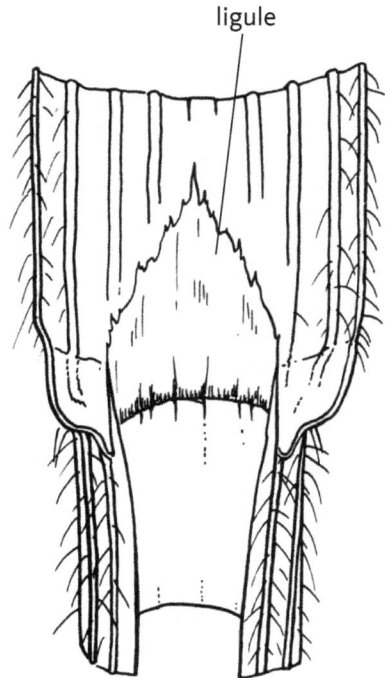

ligule

Large, acute ligule of *Bromus arizonicus*; see page 67.

Hilaria rigida
BIG GALLETA

CONDENSED KEY

Major Character Headings by Number in Key

Section A. Annual grasses

Section B. Perennial grasses

KEY NO.	CHARACTER
1	Auricle present
3	Blade glossy on lower surface
5	Bulb present
6	Flowers and blades clustered at apex
7	Blade tip very sharp
8	Ligule large, 5-18 mm long
10	Stolons present
14, 77	Rhizomes present
23	Sheath margin white
23, 25	Blade margin white
28	Letter 'M' or 'W' on blade
29	Rachis extends beyond florets
30	Culms rooting at nodes
33, 80	Internodes hairy or woolly
37	Blade margin with glandular hairs
42	Growth a mat or sod
44	Sheath closed partially or completely
46	Nodes hairy, pubescent, or woolly
74	Panicle included in sheath

HOW TO USE THE KEY (*following pages*)

First examine the plant to see if it is *annual* or *perennial*. If perennial, go to *Section B.* Identification is then made by a series of opposing choices, until a name is reached in the key.

In the short-cut method, refer first to the condensed key (above) which lists major characters with their number location in the key. For example, should the plant have stolons, go immediately to Item 10 in the key, and continue from there.Should this fail to bring you to an acceptable decision, start over at the beginning of the key. Remember, only 73 of the more common range grasses are covered by this key.

VEGETATIVE KEY

Section A - Plants Annual

1a Ligule a membrane or collar (a truncate-ciliate ligule is classified as membranous in the key) 2

1b Ligule hairy 6

    2a Sheath closed 3

    2b Sheath open 5

3a Blades 4-6 mm wide . . . . **20. *Bromus arizonicus*** ARIZONA BROME

3b Blades narrower, 1-4 mm wide 4

    4a Ligule acute-lacerate; internodes sometimes puberulent near nodes; mature plants reddish brown; chaparral and desert shrub communities at lower elevations . . . . **23. *Bromus rubens*** RED BROME

    4b Ligule obtuse-lacerate; internodes glabrous; generally at mid-elevations and higher . . . **24. *Bromus tectorum*** DOWNY CHESS, CHEATCRASS BROME

    5a Ligule truncate-ciliate, 1 mm; blades with dorsal glandular hairs; lower nodes geniculate; culms branched; awned; lower elevations . . . . **25. *Chloris virgata*** . . . . FEATHER FINGERGRASS

    5b Ligule acute-lacerate, 2-4 mm long; blades 2-5 mm wide 4-12 cm long; deserts to woodlands; lower elevations . . . . **57. *Poa bigelovii*** BIGELOW BLUEGRASS

    6a Blades needle-like, short and decumbent, 1/2-1 mm wide, 3-8 cm long; long hairs back of ligule and collar, 3-8 mm long; dry slopes and sandy washes at lower elevations . . . . . **61. *Schismus barbatus*** MEDITERRANEANGRASS

    6b Blade not needle-like 7

7a Collar and blade with glandular hairs, 2-3 mm; blade in bud folded, 1/2 mm wide, 1-2 cm long; ligule hairy, 1/2 mm; awns 1-2 cm; lower elevations . . . . **4. *Aristida adscensionis*** SIXWEEKS THREE-AWN

7b Collar without glandular hairs 8

    8a Blade margin with glandular hairs; culm elliptical; ligule hairy, 1/2 mm, blade in bud curled; stem branches; blades 1/2 mm wide, 2-7 cm long; woodlands at lower elevations . . . . **11. *Bouteloua aristidoides*** . . . . NEEDLE GRAMA

    8b Blade margin without a glandular hairs, flat, narrow and short, 1 mm wide, 2-4 cm long; culm round, ligule hairy, 1 mm; awns 1-2 mm; waste places at lower elevations . . . . **12. *Bouteloua barbata*** SIXWEEKS GRAMA

VEGETATIVE KEY
Section B - Plants Perennial

1a Auricle present, sometimes rudimentary 2

1b Auricle absent. 5

   2a Auricle prominent, clawed 3

   2b Auricle rudimentary . **2. *Agropyron cristatum*** CRESTED WHEATGRASS

3a Blades glossy below; introduced plant of lawns and pastures . . . .
   . . . . **44. *Lolium perenne*** PERENNIAL RYEGRASS

3b Blades not glossy below 4

   4a Rhizomes present; blade ribs prominent dorsally; foliage characteristically glaucous, blue-green **55. *Pascopyrum smithii*** WESTERN WHEATGRASS

   4b Rhizomes absent; blades conspicuously pubescent or rarely glabrous; tufted bunchgrass . . . . . . **31. *Elymus elymoides*** BOTTLEBRUSH SQUIRRELTAIL

5a Culm-bases thickened into bulb-like corms . . . . **73. *Zuloagaea bulbosa***
   . . . . BULB PANICUM

5b Culm-bases not bulb-like. 6

   6a Inflorescences and blades clustered at the apex, the blades exceeding the uppermost florets; plants small, 5–10 cm high; blades involute, 2–5 cm long, 1/4 mm wide . . . . **27. *Dasyochloa pulchella*** FLUFFGRASS

   6b Inflorescences and blades not clustered at apex 7

7a Blade tip pungent, stiff and sharp; common on loose sand, and in woodlands and pine forests. . . . . **50. *Muhlenbergia pungens*** SANDHILL MUHLY

7b Blade tips not sharp 8

   8a Ligule large, 5–18 mm 9

   8b Ligule smaller, usually 1/2–3 mm 10

9a Ligule acute-lacerate; culm elliptical; blades folded; large coarse bunchgrass .
   . . . . **47. *Muhlenbergia emersleyi*** BULLGRASS

9b Ligule acute-entire; culm round; blades rolled, occasionally flat; small bunchgrass. . . . . **48. *Muhlenbergia montana*** MOUNTAIN MUHLY

   10a Stolons present 11

   10b Stolons absent. 14

11a Sheaths pubescent or hirsute. 12

11b Sheaths glabrous 13

   12a Sheaths with glandular hairs . . . . . . **41. *Hopia obtusa*** VINE-MESQUITE

   12b Sheaths without glandular hairs. . . . . **26. *Cynodon dactylon***
   . . . . BERMUDAGRASS

VEGETATIVE KEY
Section B - Plants Perennial

13a Internodes woolly . . . . **15.** *Bouteloua eriopoda* BLACK GRAMA

13b Internodes glabrous. . . . . **37.** *Hilaria belangeri* CURLYMESQUITE

    14a Rhizomes present 15

    14b Rhizomes absent 23

15a Sheath closed, letter 'M' or 'W' on blade. . . . . **22.** *Bromus inermis* . . . . SMOOTH BROME

15b Sheath open 16

16a Ligule of hairs. 17

16b Ligule membranous. 18

17a Blades large, 10-12 mm wide, 20-40 cm long; ligule 1-2 mm; rhizomes massive, often flattened; fields, ditchbanks, roadsides. . . . . **66.** *Sorghum halepense* JOHNSONGRASS

17b Blades smaller, 2-3 mm wide, 1-6 cm long, frequently with salt deposits on upper surface . . . . **29.** *Distichlis spicata* DESERT SALTGRASS

    18a Blade folded, the tip boat-shaped; ligule truncate-entire . . . . 5 . . . . **9.** *Poa pratensis* KENTUCKY BLUEGRASS

    18b Blade not folded 19

19a Plant woolly, the bases coarse, semi-woody; dry areas at lower elevations . . . . **40.** *Hilaria rigida* BIG GALLETA

19b Plants not woolly. 20

    20a Blades rolled, growth mat-like . . . . **51.** *Muhlenbergia richardsonis* . . . . MAT MUHLY

    20b Blades not rolled, growth not mat-like 21

21a Ligule truncate-ciliate, 1-2 mm; blades 2-4 mm wide, 5-10 cm long, curled in the bud; culms branch at the nodes; dry, tight-soil flats, low- to mid-elevations . . . . **39.** *Hilaria mutica* TOBOSA

21b Ligule not truncate-ciliate 22

    22a Ligule truncate-lacerate, 1-2 mm; blade curled in the bud, involute, the base of blade usually flat, 2-5 mm wide, 4-12 cm long; dry flats on tight soils, mid- to high-elevations. . . . . **38.** *Hilaria jamesii* GALLETA

    22b Ligule obtuse-entire, 1-2 mm; blades 2-5 mm wide, 3-8 cm long; wet, cool sites, generally along streams at mid- to high-elevations. . . . . **3.** *Agrostis gigantea* REDTOP

23a Both sheath and blade margin white 24

23b Both sheath and blade margins not white 25

VEGETATIVE KEY
Section B - Plants Perennial

24a Blade folded, ligule acute-entire, 2–5 mm; culms branched; small bunchgrass of dry woodlands and pine forests, mid-elevations . . . . **45.** *Lycurus phleoides* WOLFTAIL

24b Blade needle-like, ligule truncate-toothed, growth mat-like . . . . **51.** *Muhlenbergia richardsonis* MAT MUHLY

25a Blade margins white 26

25b Blade margins not white 28

26a Ligule hairy . . . . **66.** *Sorghum halepense* JOHNSONGRASS

26b Ligule not hairy but membranous 27

27a Nodes woolly; ligule a collar, obtuse-lacerate 2–3 mm; hairs on back of ligule 3–5 mm; medium to large bunchgrass; dry, rocky slopes at low- to mid-elevations . . . . **10.** *Bothriochloa barbinodis* CANE BLUESTEM

27b Nodes hairy but scarcely woolly, the culms erect from a knotty swollen felty-pubescent base, 40–100 cm tall; sheaths glabrous to sparsely pilose; blades 3–5 mm wide, 8–12 cm long. **28.** *Digitaria californica* ARIZONA COTTONTOP

28a Blades with letters 'M' or 'W' . . . **22.** *Bromus inermis* SMOOTH BROME

28b Blades not as above 29

29a Rachis extends beyond florets . . . . . . **17.** *Bouteloua hirsuta* HAIRY GRAMA

29b Rachis does not extend beyond florets 30

30a Culms root at nodes 31

30b Culms not rooting at nodes 33

31a Blades with glandular hairs on upper side . . . . **37.** *Hilaria belangeri* . . . . CURLY-MESQUITE

31b Blades without glandular hairs on upper side 32

32a Internodes woolly . . . . **15.** *Bouteloua eriopoda* BLACK GRAMA

32b Internodes not woolly, the ligule ciliate. . . . . . . **27.** *Dasyochloa pulchella* . . . . FLUFFGRASS

33a Internodes hairy or woolly, the ligules hairy 34

33b Internodes neither hairy nor woolly 37

34a Ligule 1/2 mm. . . . . **15.** *Bouteloua eriopoda* BLACK GRAMA

34b Ligule larger, 1–2 mm 35

35a Blade rolled, 1 mm wide, 15–30 cm long. . . . . **30.** *Elionurus barbiculmis* . . . . WOOLSPIKE

35b Blade not rolled, wider 36

VEGETATIVE KEY
Section B - Plants Perennial

36a Blade flat, node woolly; ligule obtuse-lacerate 2–3 mm . . . .
. . . . **10. *Bothriochloa barbinodis*** CANE BLUESTEM

36b Blade folded; ligule, node and lower internode hairy . . . .
. . . . **65. *Setaria parviflora*** KNOTROOT BRISTLEGRASS

37a Blade margins, at least at the base, with glandular hairs 38

37b Blade margins without glandular hairs 42

38a Ligule membranous, 1–2 mm, collar with glandular hairs . . . .
. . . . **14. *Bouteloua curtipendula*** SIDEOATS GRAMA

38b Ligule not membranous 39

39a Culm flattened in the bud . . . . . **36. *Heteropogon contortus*** TANGLEHEAD

39b Culm round in cross section 40

40a Ligule 2–4 mm. . . . . **12. *Bouteloua barbata*** SIXWEEKS GRAMA

40b Ligule smaller 41

41a Blade in bud clasped . . **13. *Bouteloua chondrosioides*** SPRUCETOP GRAMA

41b Blade in bud curled . . . . **18. *Bouteloua repens*** SLENDER GRAMA

42a Growth a sod or mat-like 43

42b Growth not a sod or mat 44

43a Ligule hairy . . . . **16. *Bouteloua gracilis*** BLUE GRAMA

43b Ligule membranous. . . . . **53. *Muhlenbergia torreyi*** RING MUHLY

44a Sheaths closed 45

44b Sheaths open 46

45a Ligule truncate-notched; culm round. . . . . **21. *Bromus ciliatus***
. . . . FRINGED BROME

45b Ligule truncate-toothed, culm elliptical. . . . . **19. *Bromus anomalus***
. . . . NODDING BROME

46a Nodes woolly or pubescent 47

46b Nodes not woolly or pubescent 51

47a Ligule obtuse-lacerate. . . . . **10. *Bothriochloa barbinodis*** CANE BLUESTEM

47b Ligule not obtuse-lacerate 48

48a Blades folded . . . . . **65. *Setaria parviflora*** KNOTROOT BRISTLEGRASS

48b Blades not folded 49

49a Blades flat . . . . **40. *Hilaria rigida*** BIG GALLETA

49b Blades round 5 0

VEGETATIVE KEY

Section B - Plants Perennial

50a Blade small, about 1 mm wide, 2–5 cm long; ligule truncate-ciliate; blade tip very sharp to the touch; sandy soils of woodlands and pine forests . . . . **50. *Muhlenbergia pungens*** SANDHILL MUHLY

50b Blades larger, 2–4 mm wide, 8–20 cm long; sheath margin hairy; blade in bud curled; blade usually rolled; ligule small, hairy; desert grasslands and low-elevation oak woodlands . . . . **72. *Tridens muticus*** SLIM TRIDENS

51a Blades folded 52

51b Blade not folded 60

52a Growth a mat, ring-like . . . . . . **53. *Muhlenbergia torreyi*** RING MUHLY

52b Growth neither mat-like nor ring-like 53

53a Ligule large, 8–15 mm . . . . **47. *Muhlenbergia emersleyi*** BULLGRASS

53b Ligule smaller 54

54a Culm branched . . . . **45. *Lycurus phleoides*** WOLFTAIL

54b Culm not branched 55

55a Ligule shorter than 1 mm 56

55b Ligule longer than 1 mm 58

56a Ligule obtuse-lacerate . . . . **63. *Schizachyrium scoparium*** . . . . LITTLE BLUESTEM

56b Ligule not obtuse-lacerate 57

57a Ligule obtuse-toothed, about 1/2 mm; blades long, narrow, folded; woodlands and pine forests, mostly mid- to high-elevations . . . . . . **58. *Poa fendleriana*** . . . . MUTTON BLUEGRASS

57b Ligule truncate-entire, 1/4–1/2 mm; bunch or sodgrass; common in moist meadows at mid- and high-elevations . . . . **59. *Poa pratensis*** . . . . KENTUCKY BLUEGRASS

58a Ligule obtuse-entire, 1–2 mm; old growth usually flat, new growth blades folded, long, narrow, plant commonly has a reddish cast; woodlands and edges of pine forests, mid-elevations . . . . . **62. *Schizachyrium cirratum*** . . . . TEXAS BLUESTEM

58b Ligule not obtuse-entire 59

59a Ligule acute-lacerate, 2–4 mm; blade 1 mm wide, 10–20 cm long, common in chaparral, mid-elevations . . . . **58. *Poa fendleriana*** MUTTON BLUEGRASS

59b Ligule acute-entire, 2–3 mm; blade margin white; woodlands and pine forests, often where disturbed . . . **60. *Schedonnardus paniculatus*** TUMBLEGRASS

60a Blades round or needle-like in fully developed leaves 61

VEGETATIVE KEY

Section B - Plants Perennial

60b Blades flat, not round or needle-like 74

61a Ligule a collar or membranous 62

61b Ligule of hairs 70

    62a Ligule large, 5–8 mm, acute-entire; blade in bud folded, a small bunchjgrass, common in woodlands and pine forests; mid- to high-elevations .... **48. *Muhlenbergia montana*** MOUNTAIN MUHLY

    62b Ligule smaller 63

63a Ligule less than 1mm long 64

63b Ligule longer than 1 mm 68

    64a Ligule truncate-ciliate 65

    64b Ligule not truncate-ciliate 66

65a Ligule with hairs on back, 1–2 mm; blade in bud curled; short rhizomes often present; large coarse bunchgrass of alkali flats; low- to mid-elevations .... **67. *Sporobolus airoides*** ALKALI SACATON

65b Ligule with no hairs on back, truncate-ciliate, 1/4–1/2 mm; blade clasped in the bud; culms occasionally branched at nodes; small bunchgrass of dry uplands; low- to mid-elevations . **7. *Aristida purpurea*** BLUE THREE-AWN

    66a Ligule obtuse-ciliate, 1/4 mm; blade 1/2 mm wide, 25–35 cm long, needle-like, awns long, twisted, with long hairs extending to the tip .... **35. *Hesperostipa neomexicana*** NEW MEXICAN FEATHERGRASS

    66b Ligule not obtuse-ciliate 67

67a Ligule truncate-entire, 1/4 mm; blade 1 mm wide, 5–10 cm long; small bunchgrass of pine forests, mid- to upper elevations, occasionally lower.... .... **9. *Blepharoneuron tricholepis*** PINE DROPSEED

67b Ligule acute-entire, very small, 1/4–1/2 mm; blade 1/2–1 mm wide, 10–30 cm long; large bunchgrass of forests, parks and dry meadows; upper elevations. .... **34. *Festuca arizonica*** ARIZONA FESCUE

    68a Ligule lacerate 69

    68b Ligule entire, 2–3 mm; collar margin pubescent; blade 1/2 mm wide, 5–20 cm long; common in juniper-pinyon, mid-elevations .... **1. *Achnatherum hymenoides*** INDIAN RICEGRASS

69a Blade veins each side of midvein 2; blade in bud clasped; sheath margin white; large coarse bunchgrass, commonly on moist, rocky sites along intermittent streams; low- to mid-elevations .... **52. *Muhlenbergia rigens*** DEERGRASS

69b Blade veins each side of midvein 3–4; juniper and pine woodlands, mid-elevations .... **46. *Muhlenbergia curtifolia*** UTAH MUHLY

VEGETATIVE KEY
Section B - Plants Perennial

70a Blade in bud clasped 71

70b Blade in bud folded 73

71a Liguleand collar bearded; dry woodlands and pine forests at mid-elevations.
. . . . **6. *Aristida divaricata*** POVERTY THREE-AWN

71b Ligule not bearded 72

    72a Awns 2-4 cm long; culms branch freely; blade hairs on upper surface 2-3
mm . . . . **5. *Aristida californica*** SANTA RITA THREE-AWN

    72b Awns 5-8 cm long; culms unbranched; blades glabrous on upper surface
. . . . **7. *Aristida purpurea*** BLUE THREE-AWN

73a Collar bearded; blades 1/2 mm wide, 2-8 cm long; awns 2-4 mm, with
purplish cast; blades rolled, needle-like; deserts to pine woodlands, low- to
mid-elevations . . . . **7. *Aristida purpurea*** BLUE THREE-AWN

73b Collar not bearded, awns 1 long, 2 short; culm unbranched; collar with hairy
margin only; culms elliptical; dry places at low-elevations . . . .
. . . . **8. *Aristida ternipes*** SPIDERGRASS

    74a Panicle exserted, not included in the sheath 75

    74b Panicle partly or wholly covered by the sheath 87

75a Ligule of hairs for half or more its length 76

75b Ligule membranous, sometimes ciliate 82

    76a Blades sparsely pilose to pubescent, at least below 77

    76b Blades glabrous 79

77a Plants with rhizomes, forming an open to dense sod; blades narrow 78

77b Plants without rhizomes; blades 4-5 mm wide . . . . **43. *Leptochloa dubia***
. . . . GREEN SPRANGLETOP

    78a Blades rolled; salt crystals often on blades; lower culms and rhizomes
conspicuously shiny . . . . **29. *Distichlis spicata*** DESERT SALTGRASS

    78b Blades flat, the older ones often curled; salt deposits wanting; culms and
rhizomes not as above . . . . **16. *Bouteloua gracilis*** BLUE GRAMA

79a Sheaths pubescent on the margin or summit 80

79b Sheaths glabrous 81

    80a Nodes and internodes glabrous; collar margins distinctly tufted, culms
round . . . . **70. *Sporobolus giganteus*** GIANT DROPSEED

    80b Nodes and internodes sparsely hairy; collar margin not tufted; culms flat
. . . . **64. *Setaria macrostachya*** PLAINS BRISTLEGRASS

VEGETATIVE KEY
Section B - Plants Perennial

81a Culms branched, blades 12–15 cm long . . . . **33.** *Eragrostis lehmanniana*
. . . . LEHMANN LOVEGRASS
81b Culms simple, blades 45–60 cm long . . . **71.** *Sporobolus wrightii* SACATON
    82a Leaf folded in the bud; veins prominent ventrally; small bunchgrass. . . .
. . . . **42.** *Koeleria macrantha* JUNEGRASS
    82b Leaf rolled in the bud 83
83a Culms branched 84
83b. Culms not branched 86
    84a Sheaths and nodes pubescent; introduced grass . . . .
. . . . **33.** *Eragrostis lehmanniana* LEHMANN LOVEGRASS
    84b Sheaths and nodes glabrous 85
85a Wiry decumbent bunchgrass, often clambering over shrubs; blades 3–8 cm
long . . . . **49.** *Muhlenbergia porteri* BUSH MUHLY
85b Erect bunchgrass, blades 15–30 cm long . . . . . **54.** *Pappophorum vaginatum*
. . . . FEATHER PAPPUSGRASS
    86a Nodes pubescent; ligule obtuse-lacerate. . . . . **56.** *Piptochaetium pringlei*
. . . . PRINGLE NEEDLEGRASS
    86b Nodes glabrous; ligule truncate-ciliate . . . . . . **32.** *Eragrostis intermedia*
. . . . PLAINS LOVEGRASS
87a Culms branched . . . . **68.** *Sporobolus compositus* TALL DROPSEED
87b Culms not branched . . . . . . **69.** *Sporobolus cryptandrus* SAND DROPSEED

Achnatherum hymenoides
INDIAN RICEGRASS

1 *Achnatherum hymenoides* (Roemer & J.A. Schultes) Barkworth
INDIAN RICEGRASS

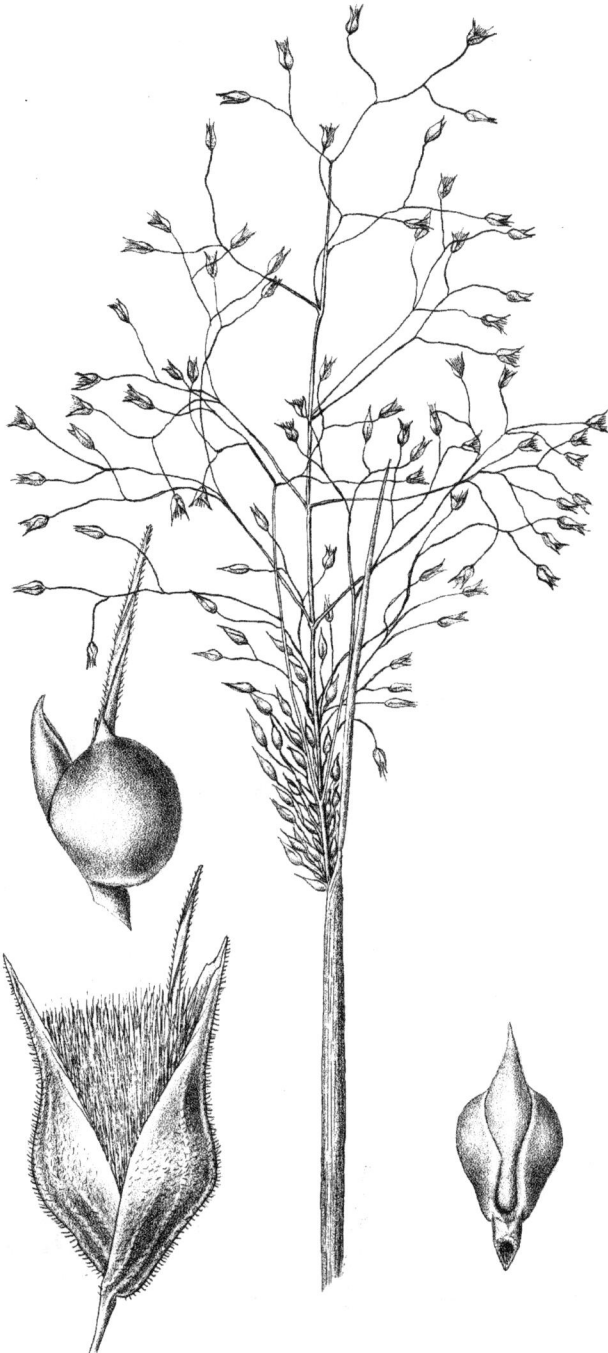

Achnatherum hymenoides (Roemer & J.A. Schultes) Barkworth
INDIAN RICEGRASS

1

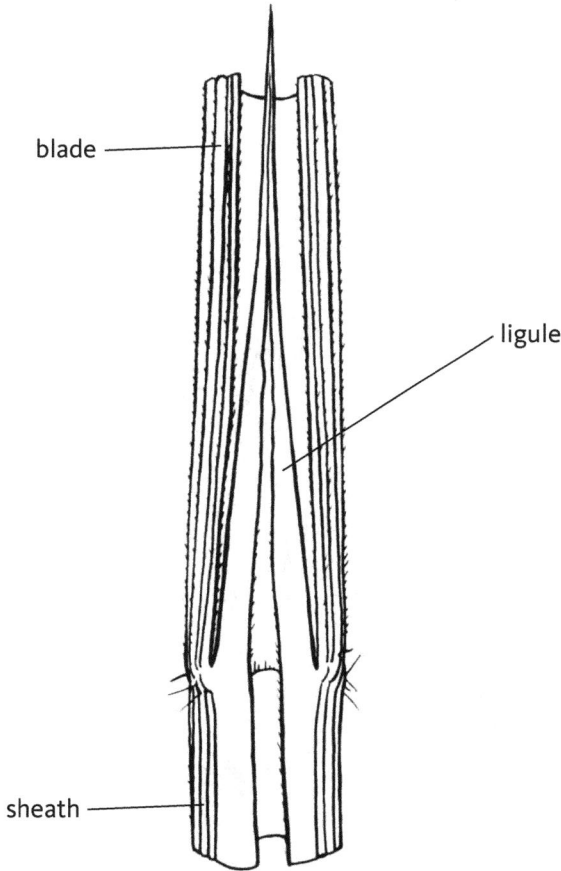

blade

ligule

sheath

SYNONYM *Oryzopsis hymenoides* (Roemer & J.A. Schultes) Ricker ex Piper

KEY CHARACTERS Small, erect bunchgrass. Sheath margin usually hairy. Ligule hyaline, acute-entire. Blade rolled, narrow, long.

VERNATION Clasped or curled. **BLADES** Rolled, drooping, narrow, pointed; rough dorsally; midrib prominent dorsally; width 1/2 mm, length 5-30 cm. **AURICLE** None. **LIGULE** Membranous, hyaline 2-3 mm, acute-entire. **COLLAR** Glabrous to pubescent. **SHEATH** Margin hairy, 1 mm. **NODE** Glabrous. **INTERNODE** Glabrous to pubescent. **ROOTS** Fibrous. **CULM** Round, not branched.

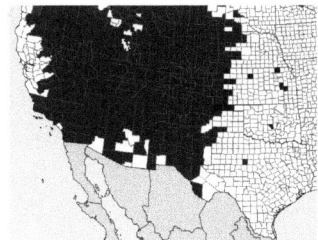

HABITAT Dry, well-drained soils.

2 *Agropyron cristatum* (L.) Gaertn.
CRESTED WHEATGRASS

Agropyron cristatum (L.) Gaertn.
CRESTED WHEATGRASS

2

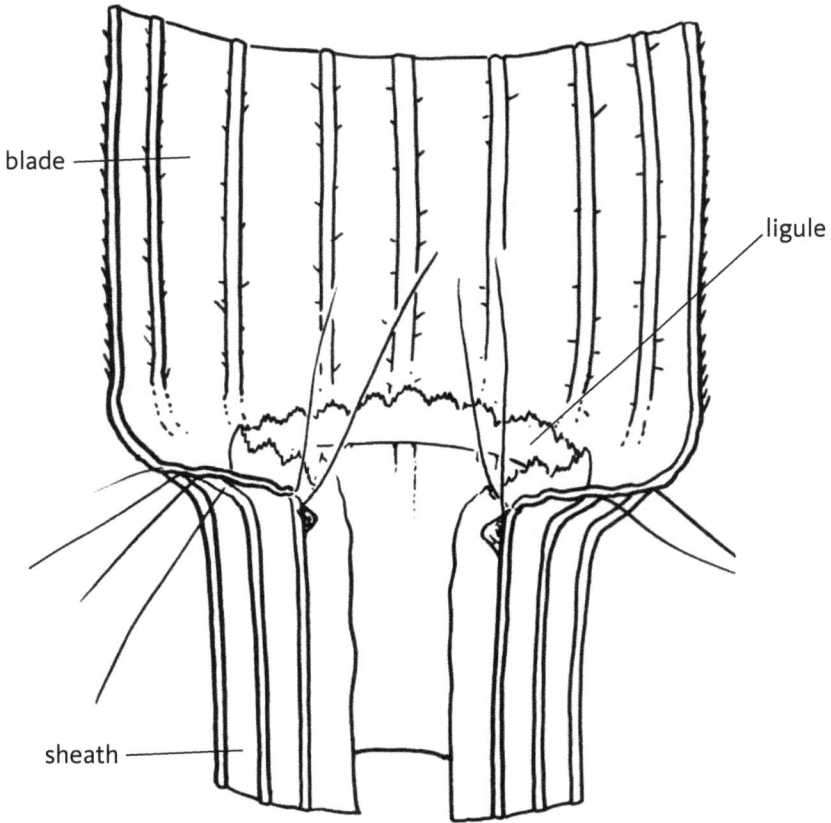

SYNONYM *Agropyron desertorum* (Fisch.) Schult.

KEY CHARACTERS Introduced, erect, medium-sized bunchgrass. Small auricle, growth in very early spring.

VERNATION Curled. **BLADES** Flat, 2–5 mm wide, 8–20 cm long; semi-erect; rough ventrally; veins each side of midrib usually three; ribs prominent dorsally, 12–18; blade margin toothed. **AURICLE** Small, often rudimentary. **LIGULE** Membranous, collar-like, 1/2 mm long. **COLLAR** Glabrous. **SHEATH** Glabrous, margin papery. **NODE** Glabrous. **INTERNODE** Glabrous. **ROOTS** Fibrous. **CULM** Round, unbranched.

HABITAT Often planted to restore productivity to areas that have been overgrazed, burned, mined, or otherwise disturbed.

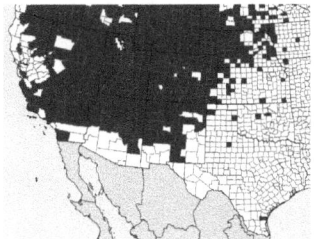

33

Agrostis gigantea Roth

REDTOP

Agrostis gigantea Roth
REDTOP

3

blade

ligule

sheath

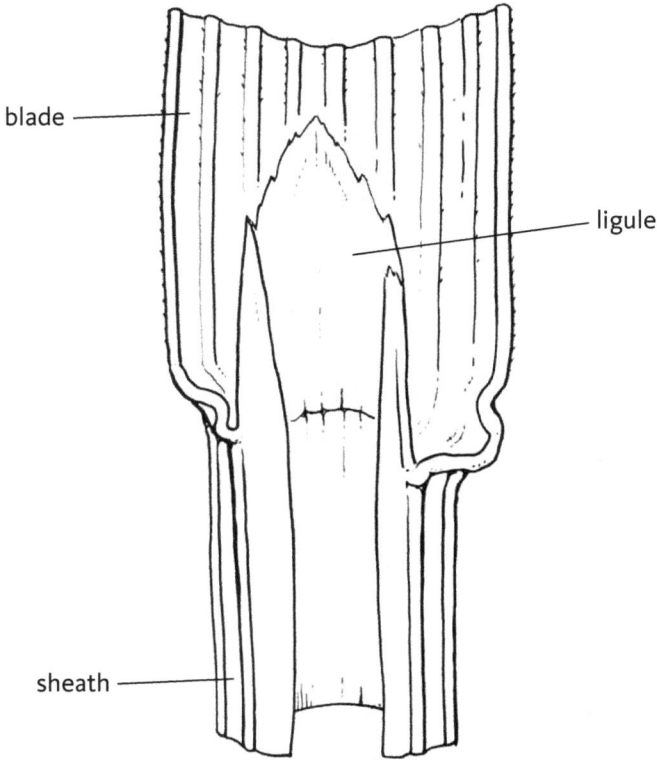

SYNONYM *Agrostis alba* L.

KEY CHARACTERS Erect, open sod-former. Ligule small, 1–3 mm; obtuse-entire, blades flat, wide tapered to point, glabrous, culm frequently reddish-purple at maturity, rhizomatous.

VERNATION Folded (also reported as curled). **BLADES** Flat, erect, narrow, pointed, wider at base; rough ventral and dorsal; veins each side midrib, 3; ribs prominent dorsally and ventrally; margin toothed; midrib prominent dorsally; 3–5 mm wide, 3–8 cm long. **AURICLE** None. **LIGULE** Membranous, small, 1–3 mm, obtuse-entire to slightly toothed. **COLLAR** Glabrous, divided. **SHEATH** Glabrous, round, frequently purplish-red. **NODE** Glabrous. **INTERNODE** Glabrous. **ROOTS** Rhizomatous. **CULM** Round, occasionally branched.

HABITAT Fields, roadsides, ditches, and other disturbed habitats, mostly at lower elevations; can become weedy.

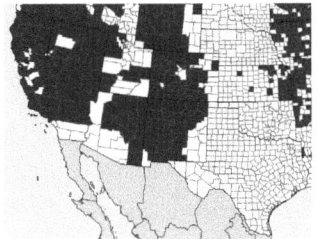

35

Aristida adscensionis L.

SIXWEEKS THREE-AWN

Aristida adscensionis L.

SIXWEEKS THREE-AWN

4

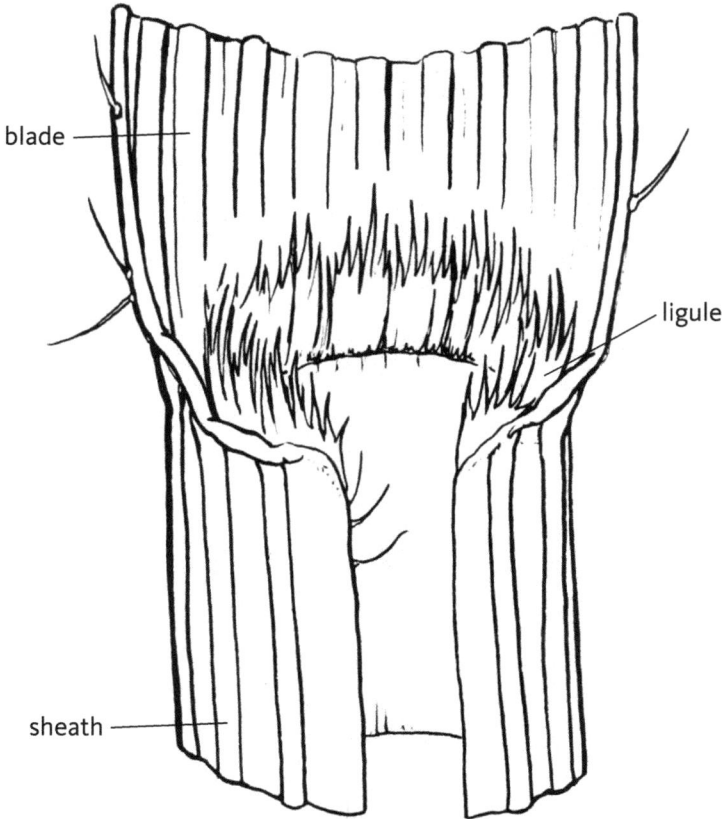

KEY CHARACTERS Small annual, erect bunchgrass. Collar with glandular hairs. Ligule hairy, very small, 1/2 mm. Roots superficial. Blades narrow, short, usually margin glandular hairs. At maturity usually blown away. Awns, 1–2 cm.

VERNATION Folded. BLADES Flat, narrow, short; glabrous, soft; veins each side of midrib, 3–4; ribs not prominent; margin glandular hairs at base; midrib not prominent; 1/2–1 mm wide, 1–2½ cm long. AURICLE None. LIGULE Hairy, small 1/2 mm, long hairs, 2–3 mm. COLLAR Margin hairy, glandular, 2–3 mm. SHEATH Occasional hairs, margin papery. NODE Glabrous. INTERNODE Glabrous. ROOTS Fibrous, superficial. CULM Round, frequently branched.

HABITAT Waste ground, roadsides, degraded rangelands and dry hillsides, often in sandy soils. It is associated with woodland, prairie, and desert shrub communities.

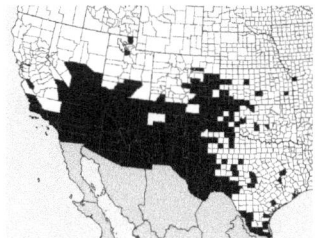

Aristida californica Thurb.
SANTA RITA THREE-AWN

Aristida californica Thurb.
SANTA RITA THREE-AWN

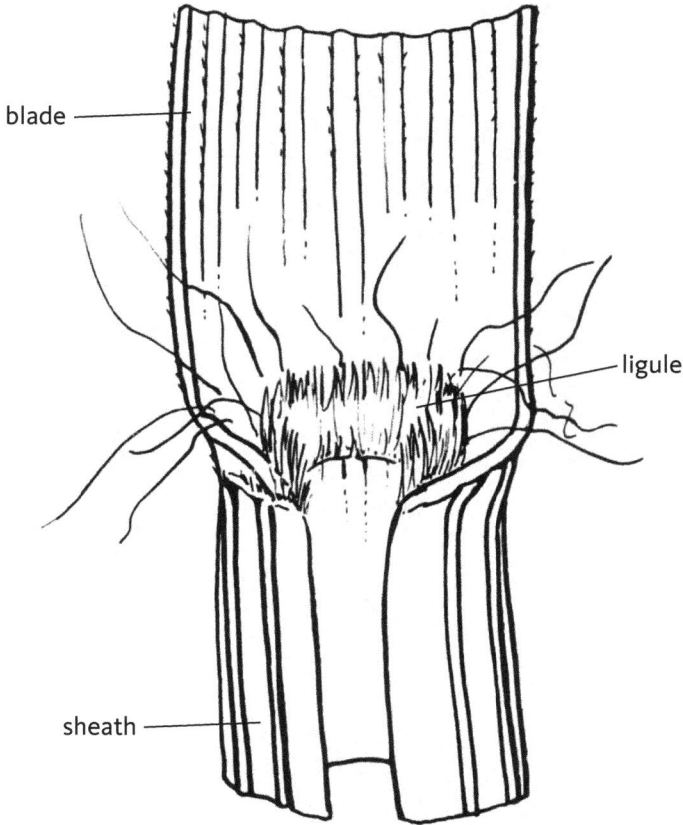

blade

ligule

sheath

....

SYNONYM *Aristida glabrata* (Vasey) A.S. Hitchc.

....

KEY CHARACTERS Erect, small bunchgrass. Blades long, narrow, usually rolled, with scattered hairs, 2–3 mm dorsally. Ligule hairy, small, and with scattered hairs, 2–3 mm. Awns 2–4 cm. Culms branched.

....

VERNATION clasped. **BLADES** Hairy ventrally on lower 1/4, 2–3 mm; veins each side of midrib, 3; ribs prominent ventrally and dorsally; margin glabrous; midrib not prominent; 1 mm wide, 5–18 cm long. **AURICLE** None. **LIGULE** Hairy, small 1/2–1 mm, occasional hairs 2–3 mm. **COLLAR** Glabrous, divided. **SHEATH** Glabrous, veined, margin papery. **NODE** Glabrous. **INTERNODE** Glabrous. **ROOTS** Fibrous. **CULM** Elliptical, strongly branched.

....

HABITAT Dry, sandy to rocky plains; sand dunes.

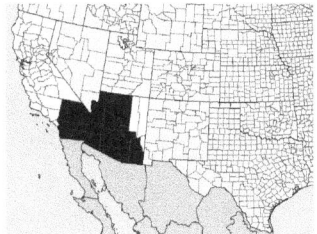

Aristida divaricata Humb. & Bonpl. ex Willd.

POVERTY THREE-AWN

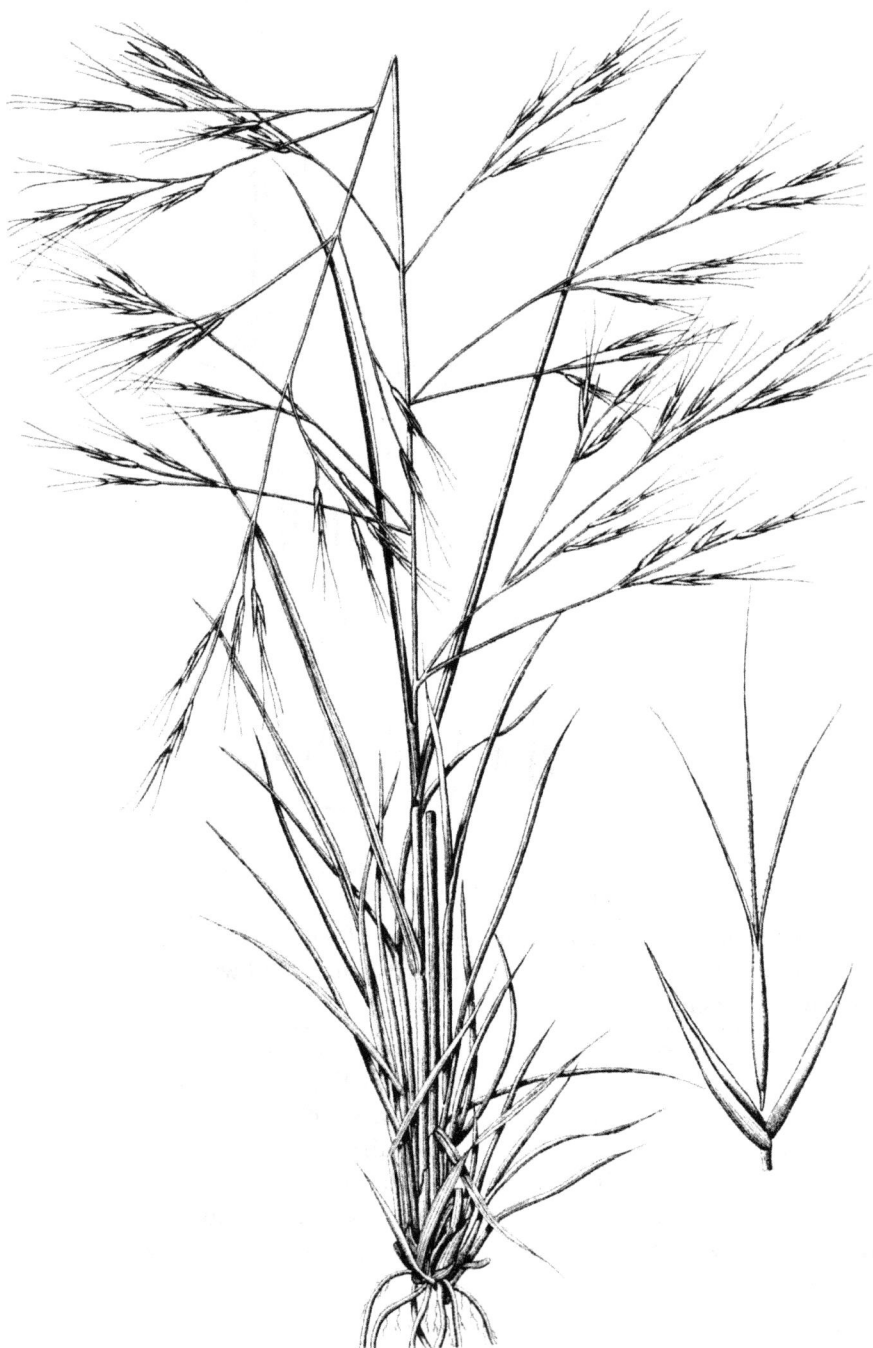

Aristida divaricata Humb. & Bonpl. ex Willd.

POVERTY THREE-AWN

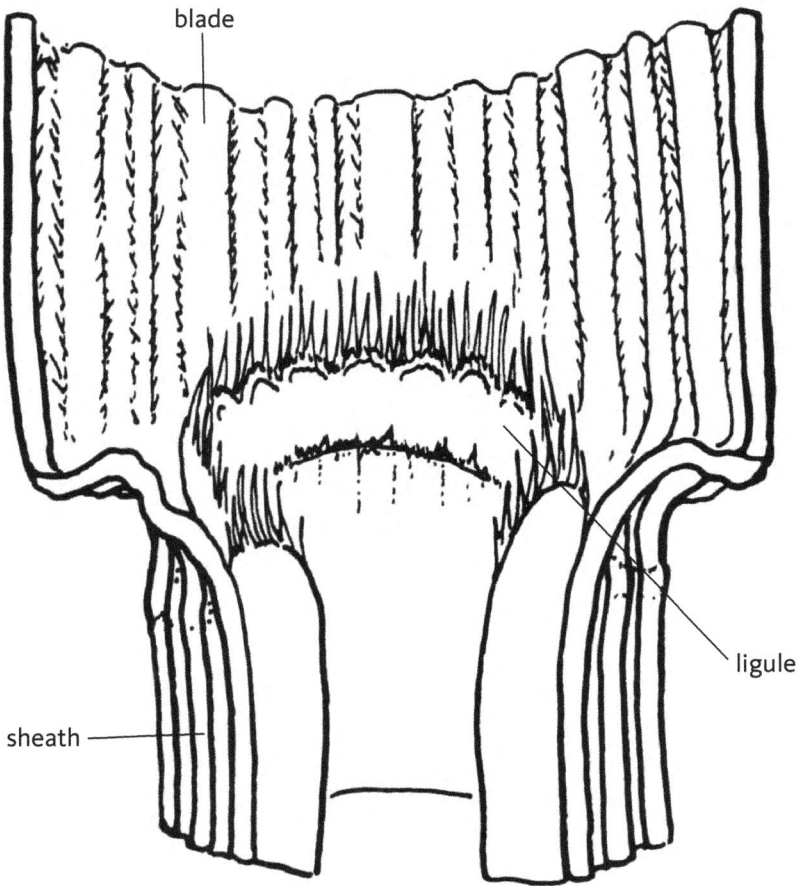

KEY CHARACTERS Small, erect bunchgrass. Collar and ligule bearded. Blades narrow, rolled, thread-like. Awns short 10–15 mm.

VERNATION clasped. **BLADES** Rolled, erect, narrow, pointed; smooth; ribs prominent ventrally; margin toothed; midrib not prominent; 1/2–1 mm wide, 5–15, occasionally 25 cm long. **AURICLE** None. **LIGULE** Hairy, small 1/2–1 mm. **COLLAR** Bearded, 1–2 mm. **SHEATH** Glabrous, veined. **NODE** Glabrous. **INTERNODE** Glabrous. **ROOTS** Fibrous. **CULM** Elliptical, not branched.

HABITAT Dry hills and plains, especially in pinyon-juniper grasslands.

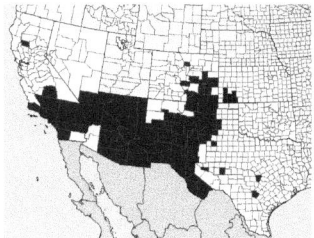

Aristida purpurea Nutt.

BLUE THREE-AWN

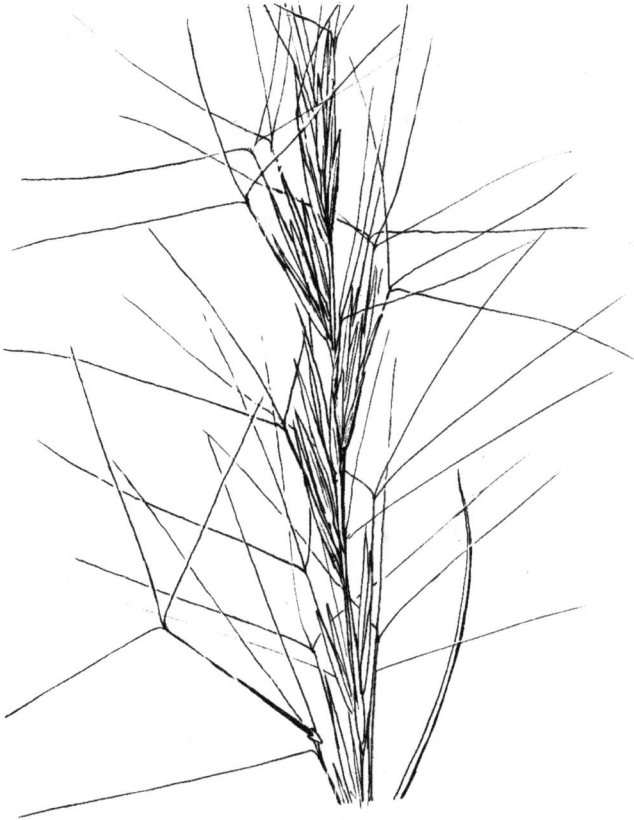

Aristida purpurea Nutt.

BLUE THREE-AWN

blade

ligule

sheath

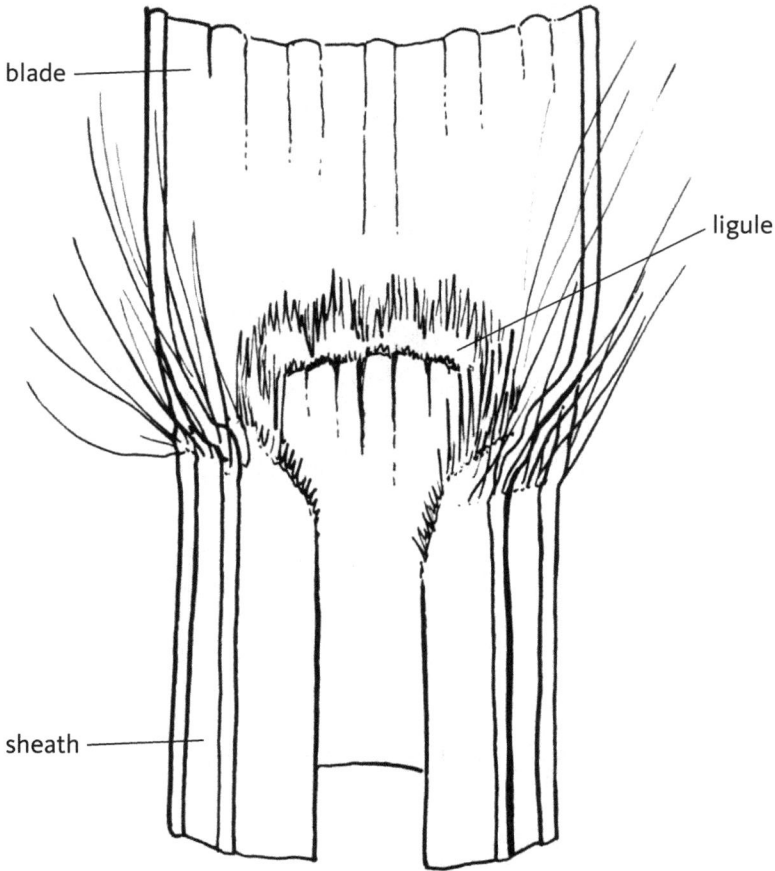

SYNONYM *Aristida glauca* (Nees) Walp.

KEY CHARACTERS Erect, medium bunchgrass. Blades rolled, narrow, long, pointed. Culms usually branched. Ligule ciliate, short. Awns 2–4 cm long.

VERNATION Clasped. **BLADES** Rolled, narrow, pointed; usually glabrous, stiff; veins and ribs indistinct; margin smooth; midrib prominent; 1/2 mm wide, 15–25 cm long. **AURICLE** None. **LIGULE** tiny, 1/4–1/2 mm, truncate-ciliate. **COLLAR** Ciliate front, margin divided. **SHEATH** Round, glabrous. **NODE** Glabrous. **INTERNODE** Glabrous. **ROOTS** Fibrous. **CULM** Round, usually branched.

HABITAT Sandy or rocky slopes and plains; barren soils.

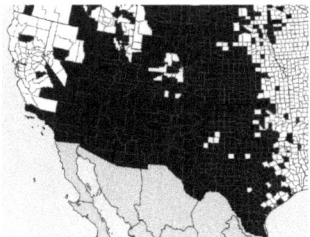

Aristida ternipes Cav.
SPIDERGRASS

Aristida ternipes Cav.
SPIDERGRASS

blade

ligule

sheath

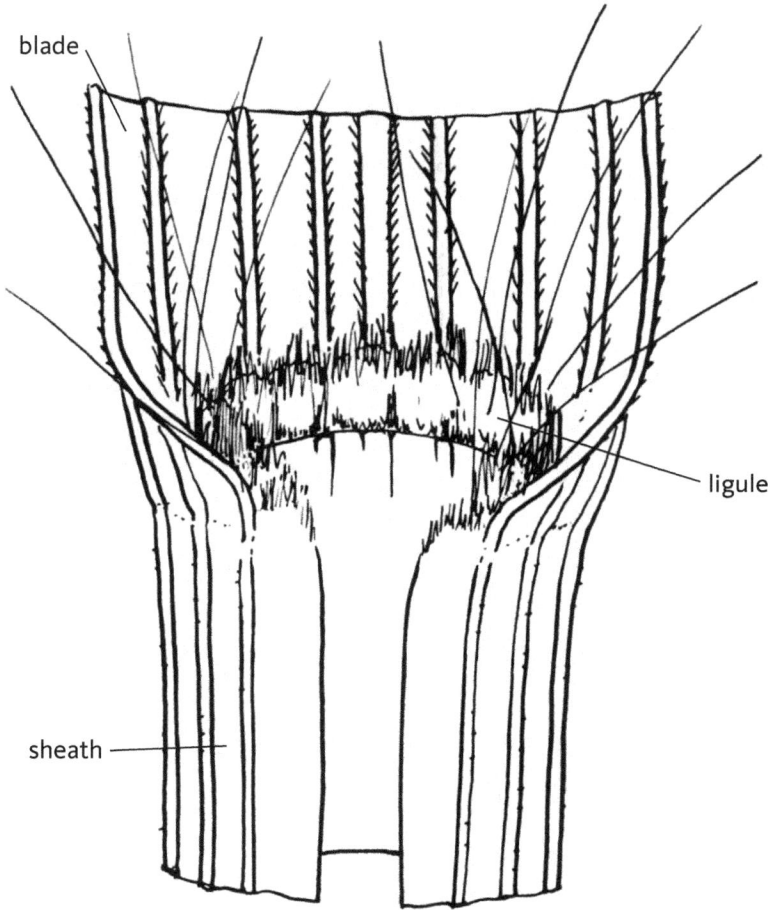

....
KEY CHARACTERS Small, erect bunchgrass. Blades flat at base, narrow, long, ribbed.
Ligule hairy, small. Collar margin hairy. Awns, 1 long and 2 short.
....
VERNATION Folded. **BLADES** Flat at base, tip rolled; rough, occasional hairs dorsally;
veins each side of midrib, 2; ribs prominent ventrally and dorsally; margin toothed;
midrib not prominent; 2 mm wide, 5-20 cm. long.
AURICLE None. **LIGULE** Hairy, small 1 mm. **COLLAR**
Margin hairy, 1 mm, divided. **SHEATH** Glabrous,
margin papery. **NODE** Glabrous. **INTERNODE** Glabrous.
ROOTS Fibrous. **CULM** Elliptical, occasionally
branched.
....
HABITAT Dry slopes and plains; roadsides.

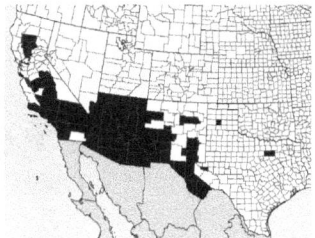

9 *Blepharoneuron tricholepis* (Torr.) Nash
PINE DROPSEED

Blepharoneuron tricholepis (Torr.) Nash

PINE DROPSEED

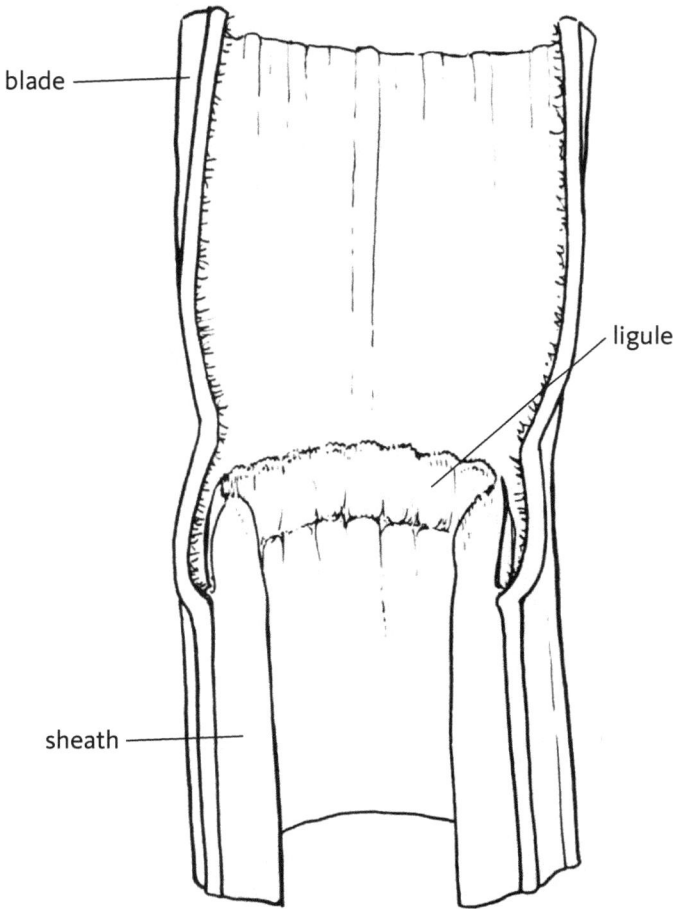

blade

ligule

sheath

KEY CHARACTERS Erect, small bunchgrass. Blades few, basal, long, narrow, usually rolled. Ligule small, truncate-entire.

VERNATION Folded. **BLADES** Usually rolled, narrow, pointed; rough, stiff, ribs indistinct; margin toothed; midrib not prominent; 1 mm wide, 5-10 cm long. **AURICLE** None. **LIGULE** Membranous, very small, 1 mm, truncate-entire. **COLLAR** Glabrous. **SHEATH** Elliptical, glabrous, veined. **NODE** Glabrous. **INTERNODE** Glabrous. **ROOTS** Fibrous. **CULM** Round, not branched.

HABITAT Dry, rocky to sandy slopes, dry meadows, open woods in pine-oak-madrone forests.

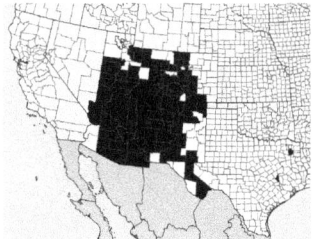

Bothriochloa barbinodis (Lag.) Herter

CANE BLUESTEM

Bothriochloa barbinodis (Lag.) Herter
CANE BLUESTEM

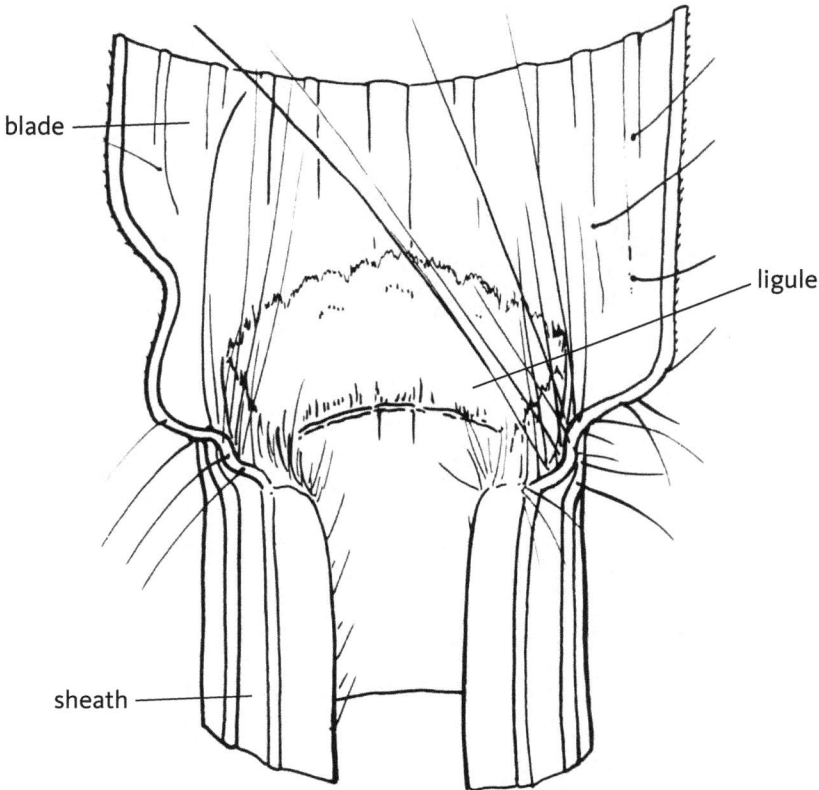

10

SYNONYM *Andropogon barbinodis* Lag.

KEY CHARACTERS Small, erect bunchgrass. Blade margin white; nodes woolly, culms large, usually branched. Mature plants usually reddish. Panicle large, cotton-like, awned.

VERNATION Curled. BLADES Flat, narrow, pointed; reddish; rough ventrally and dorsally; margin white, toothed; midrib prominent dorsally; 2–5 mm wide, 8–20 cm long. AURICLE None. LIGULE Membranous, 2–3 mm, obtuse-lacerate, occasionally long hairs on back, 3–5 mm. COLLAR Smooth, divided. SHEATH Round, reddish, smooth, hyaline margin, veined. NODE Woolly. INTERNODE Glabrous. ROOTS Fibrous. CULM Round, usually branched, reddish.

HABITAT Roadsides, drainages, gravelly slopes in desert grasslands.

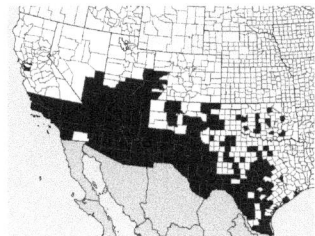

49

11 *Bouteloua aristidoides* (Kunth) Griseb.

NEEDLE GRAMA

Bouteloua aristidoides (Kunth) Griseb.
NEEDLE GRAMA

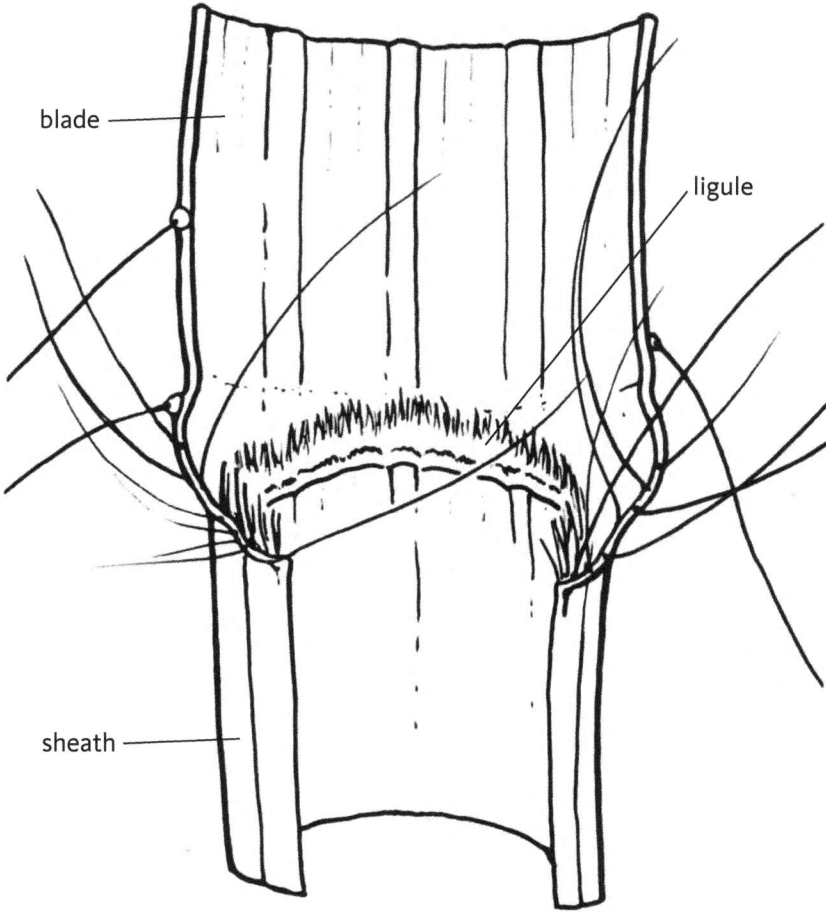

KEY CHARACTERS Small annual, semi-erect bunchgrass. Ligule hairy. Blade margin with glandular hairs. Stems branched. Awns short.

VERNATION Curled. **BLADES** Flat, erect, narrow, pointed; blade hairy near collar; veins each side of midrib, 1-2; ribs indistinct; margin with grandular hairs near ligule, 2-4 mm; midrib prominent ventrally; 1-2 mm wide, 2-7 cm long. **AURICLE** None. **LIGULE** Hairy, small 1/2 mm. **COLLAR** Glabrous. **SHEATH** Glabrous, veined, elliptical. **NODE** Glabrous. **INTERNODE** Glabrous. **ROOTS** Fibrous. **CULM** Elliptical, branched.

HABITAT Dry mesas, plains, and washes.

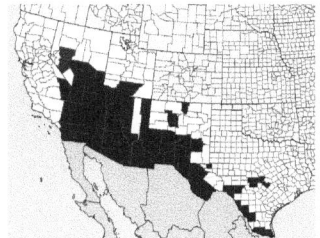

51

12

Bouteloua barbata Lag.
SIXWEEKS GRAMA

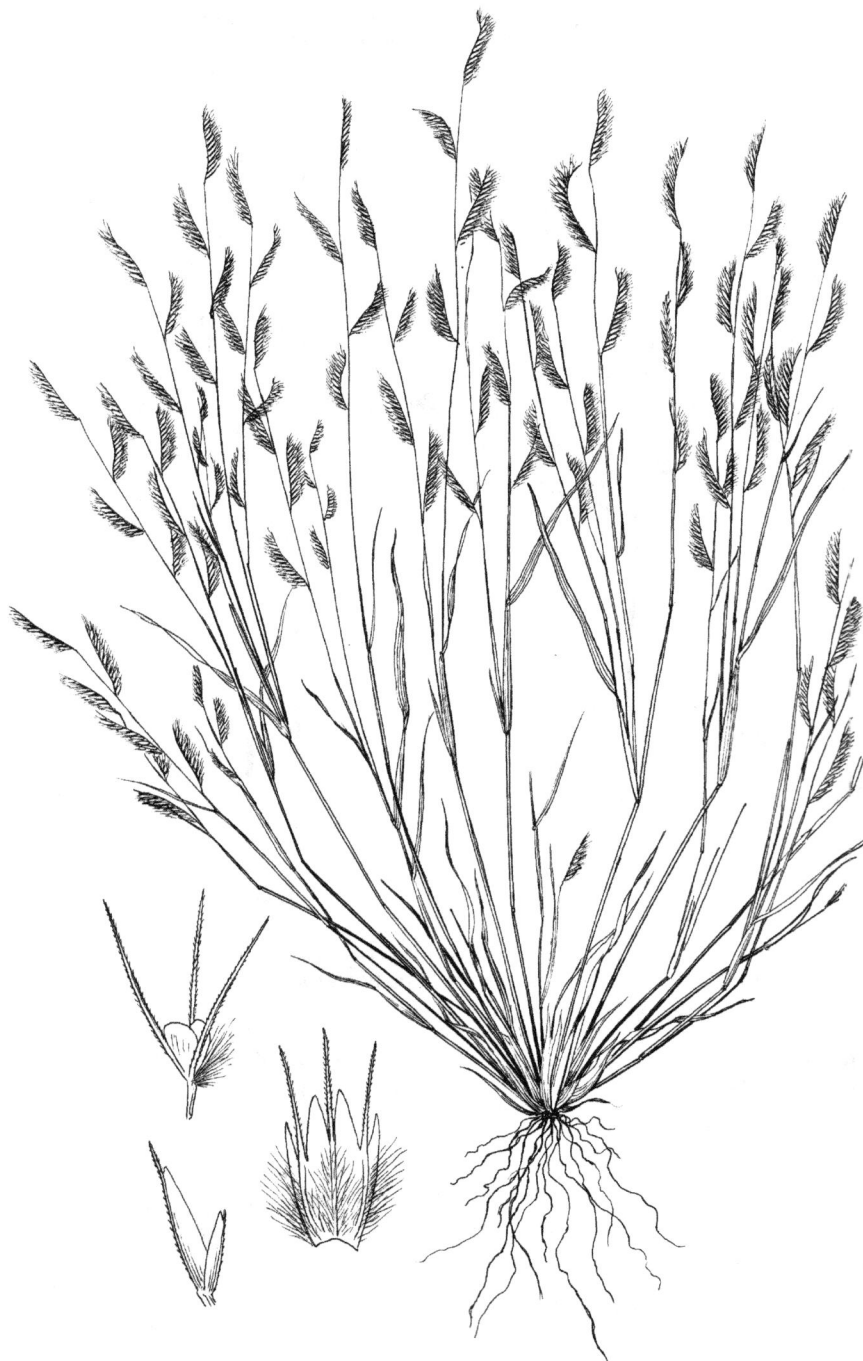

Bouteloua barbata Lag.
SIXWEEKS GRAMA

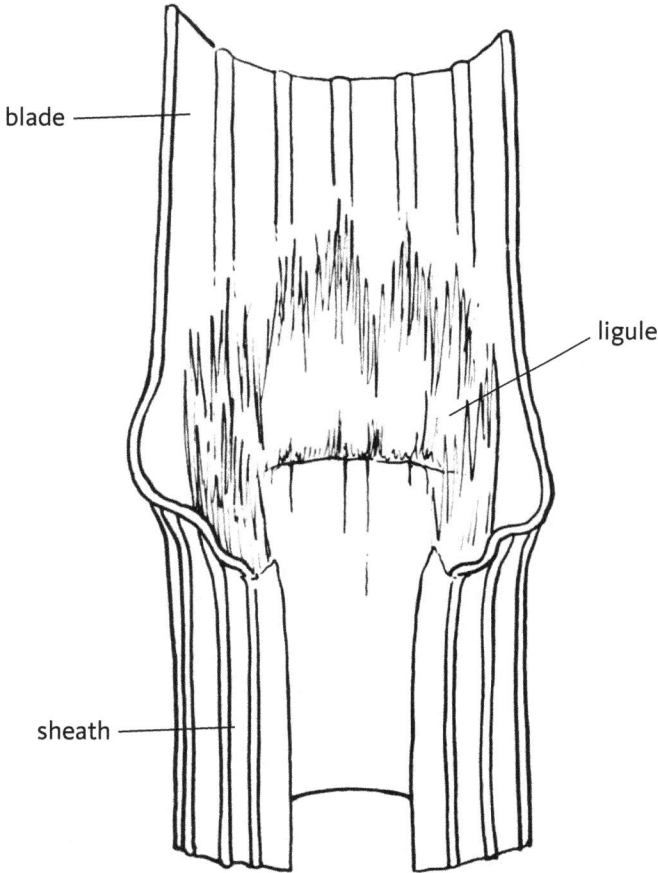

KEY CHARACTERS Small annual, erect bunchgrass. Roots superficial, blows away at maturity. Ligule hairy. Blades flat, narrow, short. Sheath margin papery. Awns short.

VERNATION Curled. **BLADES** Flat, narrow, short; glabrous; veins each side of midrib, 2; ribs not prominent; margin glabrous; midrib not prominent; 1 mm wide, 2–4 cm long. **AURICLE** None. **LIGULE** Hairy, small 1 mm. **COLLAR** Glabrous. **SHEATH** Glabrous, veined, margin papery. **NODE** Glabrous. **INTERNODE** Glabrous. **ROOTS** Fibrous. **CULM** Branched, round.

HABITAT Dry, sandy or rocky slopes; sandy flats.

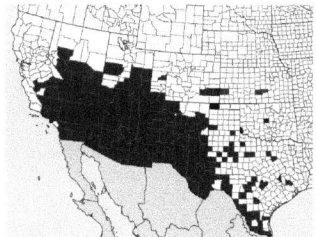

13 *Bouteloua chondrosioides* (Kunth) Benth. ex S. Wats.
SPRUCETOP GRAMA

Bouteloua chondrosioides (Kunth) Benth. ex S. Wats.
SPRUCETOP GRAMA

13

blade

ligule

sheath

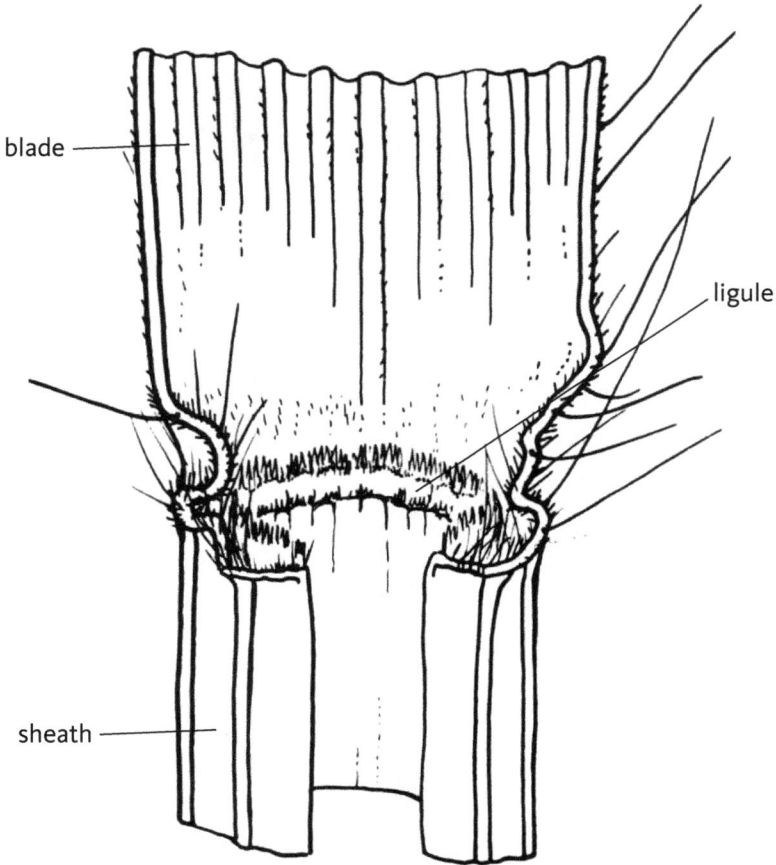

KEY CHARACTERS Small, erect bunchgrass. Glandular hairs on blade margin, dorsally and occasionally ventrally. Ligule hairy, small, 1/4 mm. Spikelet villous, short and broad.

VERNATION Clasped. **BLADES** Flat, drooping, narrow, pointed; occasionally hairy ventrally and dorsally; veins each side of midrib, 4; ribs numerous, prominent ventrally and dorsally; margin with glandular hairs, midrib prominent ventrally; 2 mm wide, 5-15 cm long. **LIGULE** Hairy, 1/4 mm. **COLLAR** Hairy dorsally, margin divided. **SHEATH** Round, veined, glabrous. **NODE** Glabrous. **INTERNODE** Glabrous. **ROOTS** Fibrous. **CULM** Round, not branched.

HABITAT Dry, rocky slopes; grassy plateaus.

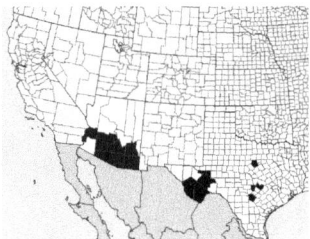

Bouteloua curtipendula (Michx.) Torr.
SIDEOATS GRAMA

Bouteloua curtipendula (Michx.) Torr.
SIDEOATS GRAMA

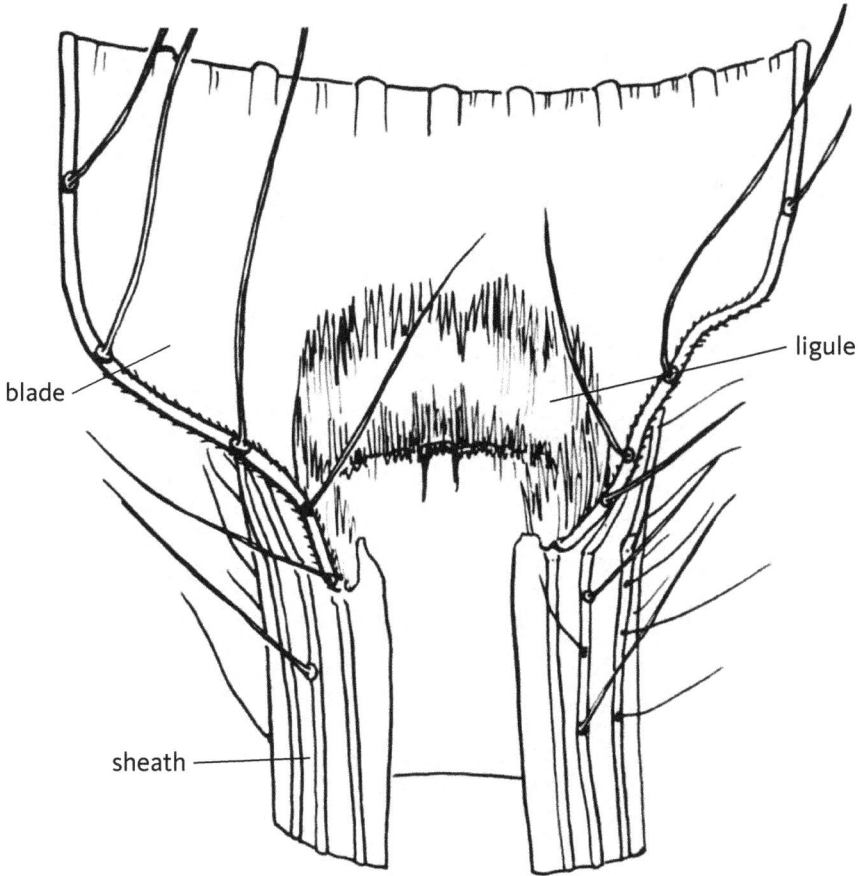

blade

ligule

sheath

KEY CHARACTERS Large, erect bunchgrass. Blade and collar margins with conspicuous glandular hairs. Rachis zigzag. Mature plants reddish-brown.

VERNATION Curled. **BLADES** Flat, long, drooping, narrow, pointed; rough dorsally; veins each side of midrib, 2-3; ribs not prominent; margin toothed, glandular hairs; midrib not prominent; 2-3 mm wide, 5-20 cm long. **AURICLE** None. **LIGULE** Membranous, small, 1 mm, truncate-lacerate. **COLLAR** Hairy margin and occasionally glandular. **SHEATH** Glabrous, veined, margin papery. **NODE** Glabrous. **INTERNODE** Glabrous. **ROOTS** Short rhizomes. **CULM** Elliptical to round, occasionally branched.

HABITAT Common, often dominant or co-dominant grass in open, sometimes moist to wet grasslands.

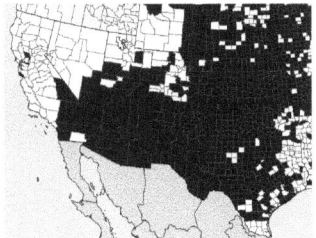

Bouteloua eriopoda (Torr.) Torr.
BLACK GRAMA

Bouteloua eriopoda (Torr.) Torr.
BLACK GRAMA

blade

ligule

sheath

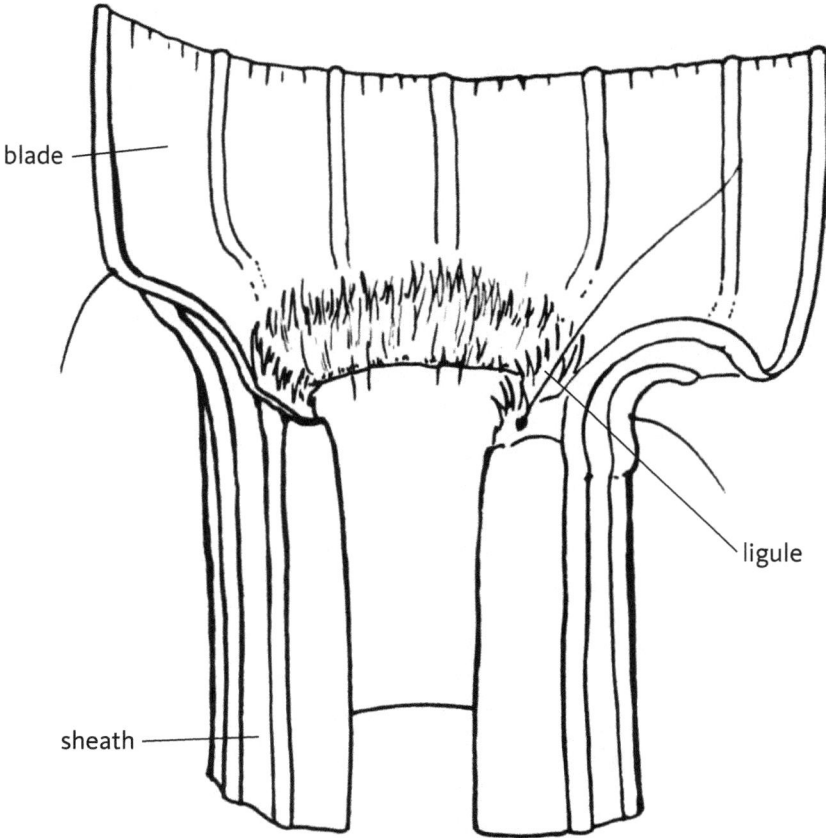

KEY CHARACTERS Erect to generally decumbent. Old plants surrounded by new plants connected with stolons. Blade margin glandular hairs. Internodes canescent. Culm roots at nodes and produces new plants. Lower stems remain green year long.

VERNATION Curled. **BLADES** Flat, narrow, drooping, pointed; soft, hairy ventrally; veins each side of midrib, usually 2; ribs not prominent; margin glandular hairy, 3–5 mm, midrib not prominent; 1–2 mm wide, 3–8 cm long. **AURICLE** None. **LIGULE** Hairy, small, 1/2 mm. **COLLAR** Glabrous or with scattered hairs. **SHEATH** Round, glabrous. **NODE** Sparsely pubescent to glabrous. **INTERNODE** Woolly. **ROOTS** Fibrous. **CULM** Round, branch at node, woolly.

HABITAT Dry plains, foothills, and open forested slopes, often in shrubby habitats; waste ground. Highly palatable, decreasing under heavy grazing.

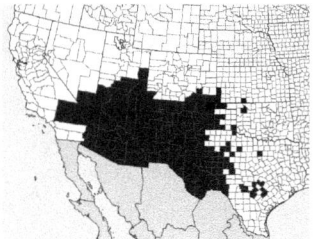

16 *Bouteloua gracilis* (Willd. ex Kunth) Lag. ex Griffiths
BLUE GRAMA

Bouteloua gracilis (Willd. ex Kunth) Lag. ex Griffiths

BLUE GRAMA

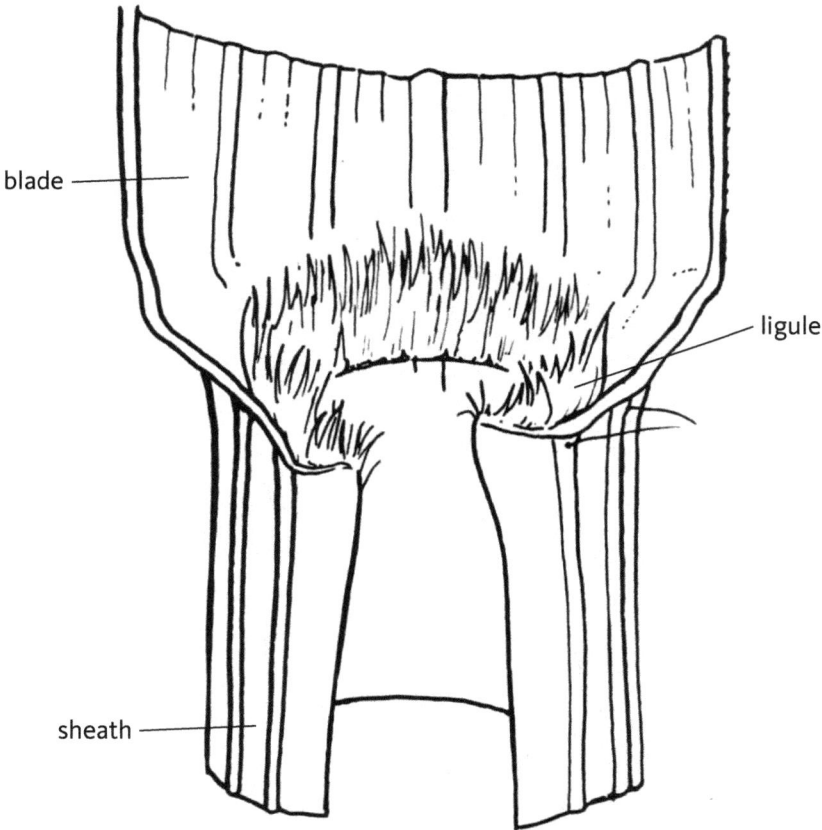

blade

ligule

sheath

KEY CHARACTERS Erect, open sod- or mat-former. Blade in bud clasped. Blades flat, narrow, long drooping, glabrous; old-growth blades curled.

VERNATION Clasped. **BLADES** Flat, drooping, narrow, pointed; rough ventrally, veins each side of midrib usually 2; veins prominent ventrally and dorsally, margin toothed; sparsely pilose below; 2–3 mm wide, 5–15 cm long. **AURICLE** None. **LIGULE** Small, 1/2 mm, mostly of hairs. **COLLAR** Glabrous (smooth, not pubescent or hairy). **SHEATH** Round, glabrous, veined. **NODE** Glabrous. **INTERNODE** Glabrous. **ROOTS** Fibrous, short rhizomes. **CULM** Round, not branched.

HABITAT May form pure stands in mixed prairie associations and disturbed habitats, usually on rocky or clay soils.

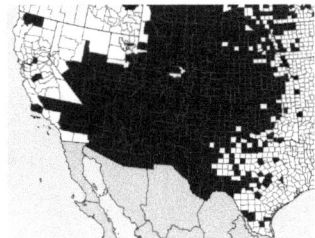

Bouteloua hirsuta Lag.
HAIRY GRAMA

Bouteloua hirsuta Lag.
HAIRY GRAMA

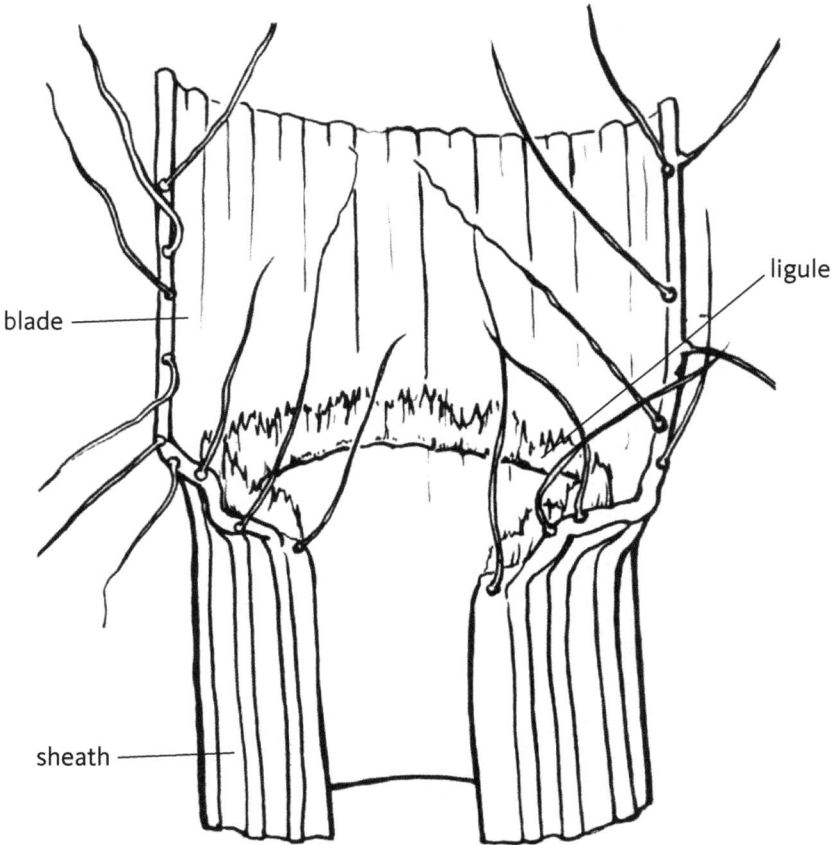

KEY CHARACTERS Small, erect bunchgrass. Blades margin with glandular hairs. Rachis extends beyond terminal spikelet. Collar margin with glandular hairs.

VERNATION Clasped. **BLADES** Flat, drooping, narrow, pointed; rough, hairy dorsally; veins each side of midrib, 3–4 ribs indistinct; margin glandular hairs; midrib not prominent; 1–3 mm wide, 3–10 cm long. **AURICLE** None. **LIGULE** Membranous, small, 1/4 mm, with occasional marginal hairs; truncate-ciliate. **COLLAR** Hairy with glandular marginal hairs. **SHEATH** Veined, round with papery margin. **NODE** Glabrous. **INTERNODE** Glabrous. **ROOTS** Fibrous. **CULM** Not branched, round.

HABITAT Open plains to shaded openings in woods and shrublands; soils well-drained, often rocky.

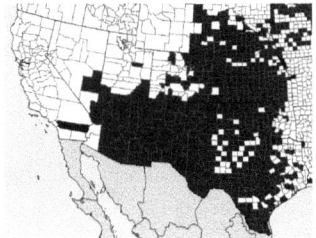

Bouteloua repens (Kunth) Scribn.

SLENDER GRAMA

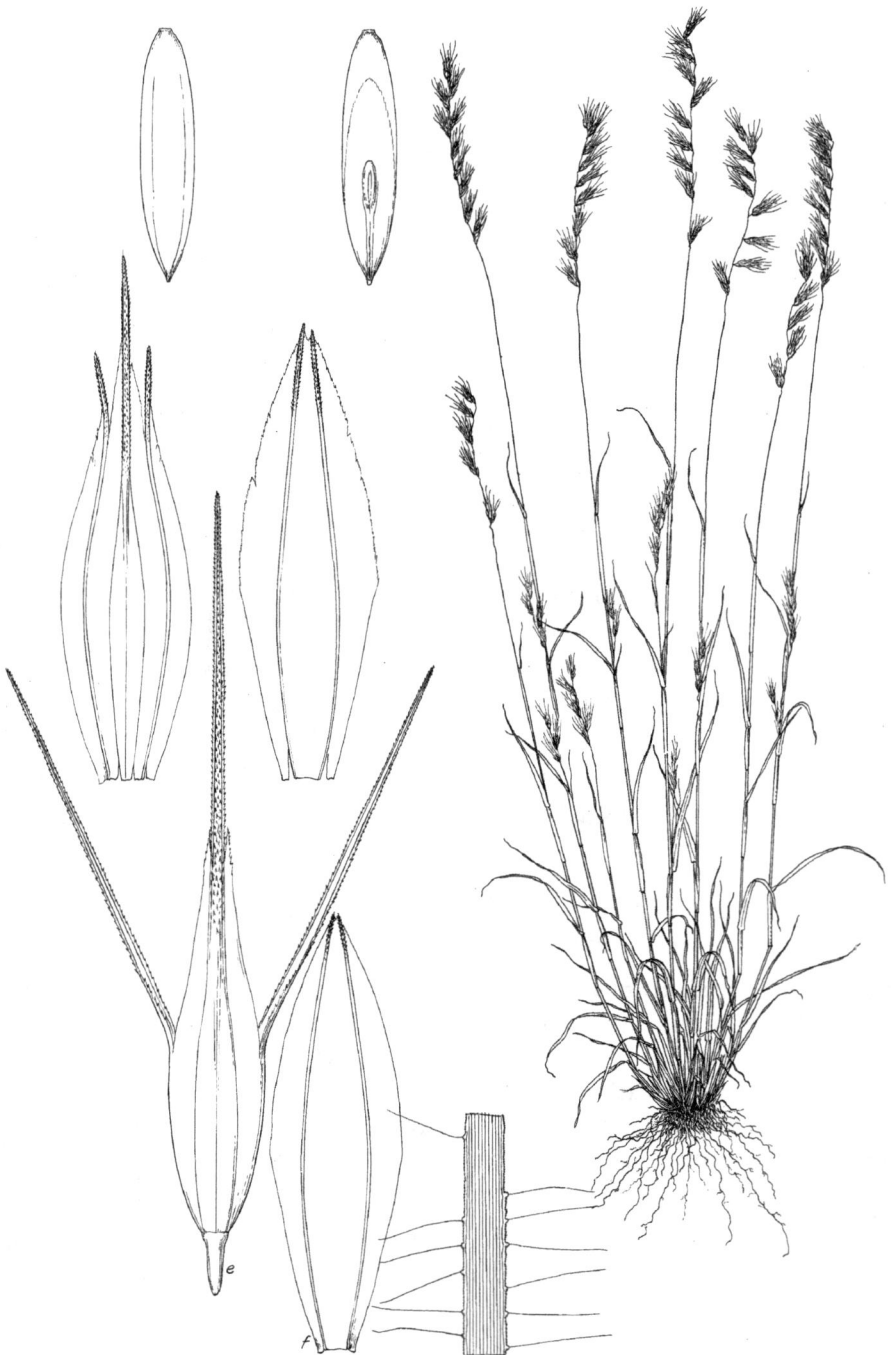

Bouteloua repens (Kunth) Scribn.
SLENDER GRAMA

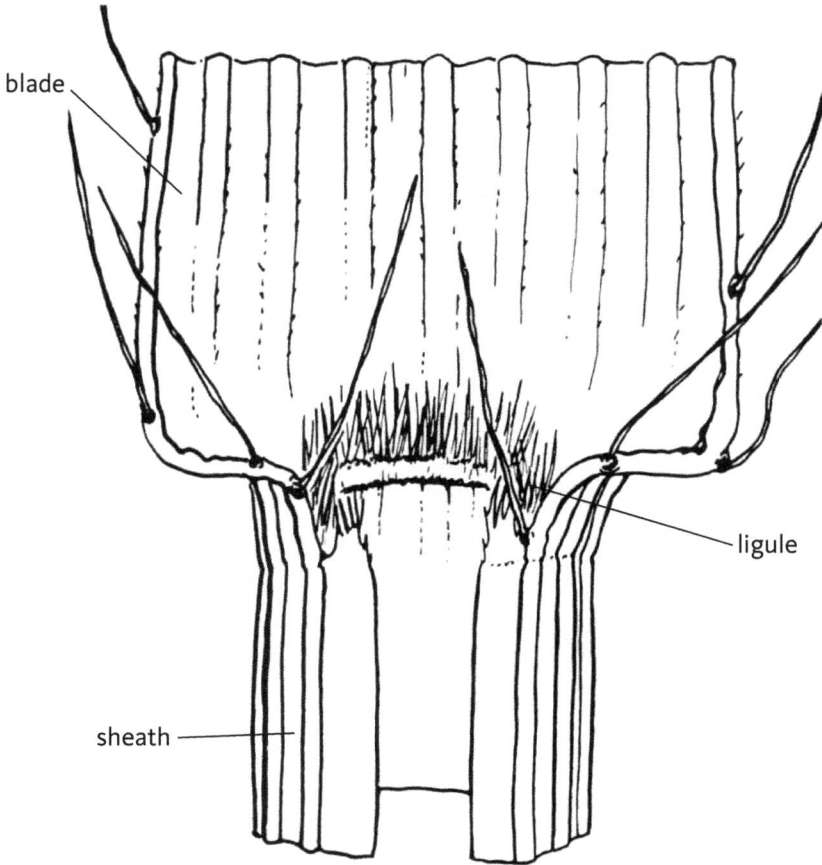

SYNONYM *Bouteloua filiformis* (Fourn.) Griffiths

KEY CHARACTERS Small, erect bunchgrass. Blade margin with glandular hairs. Ligule ciliate, small, 2/3 mm. Blades narrow, long.

VERNATION Curled. BLADES Flat, narrow, pointed; margin rough, glandular hairs; veins each side of midrib, 2–3; ribs prominent (with lens) ventrally and dorsally, midrib prominent ventrally; 1–2 mm wide, 5–8 cm long. AURICLE None. LIGULE Hairy, small 1/2 mm, ciliate. COLLAR Margin hairy. SHEATH Glabrous, margin papery. NODE Glabrous. INTERNODE Glabrous. ROOTS Fibrous. CULM Elliptical, not branched.

HABITAT Open, usually hilly terrain, soils various.

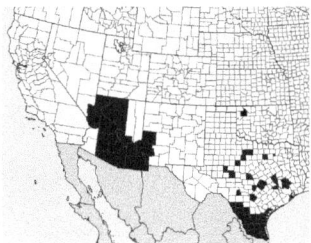

19 *Bromus anomalus* Rupr. ex Fourn.

NODDING BROME

Bromus anomalus Rupr. ex Fourn.

NODDING BROME

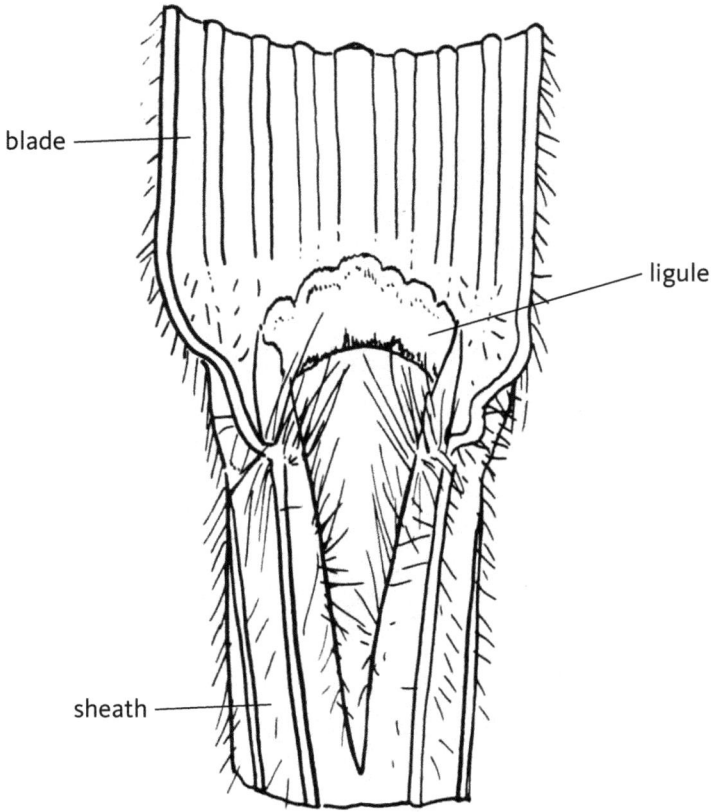

KEY CHARACTERS Small, erect bunchgrass. Sheath usually closed completely. Blades flat, wide, long, drooping; blade midrib prominent dorsally and ventrally. Blades and sheath with prominent ribs. Ligule membranous, 1/2 mm, truncate-toothed.

VERNATION Curled. **BLADES** Flat, wide, long, drooping; occasionally hairy dorsally, 2-4 mm; veins each side of midrib, 3; ribs prominent ventrally and dorsally; margin toothed; midrib prominent dorsally and ventrally; 2-8 mm. wide, 15-40 cm long. **AURICLE** None. **LIGULE** Membranous, small 1/2 mm, truncate-toothed. **COLLAR** Glabrous, divided. **SHEATH** Ribbed, generally closed, glabrous to hairy on margin, the lower sheaths sometimes retrorsely hispid. **NODE** Glabrous. **INTERNODE** Glabrous. **ROOTS** Fibrous. **CULM** Elliptical, not branched.

HABITAT Rocky slopes.

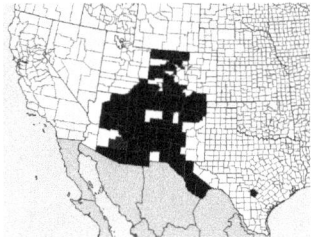

ARIZONA BROME

ARIZONA BROME

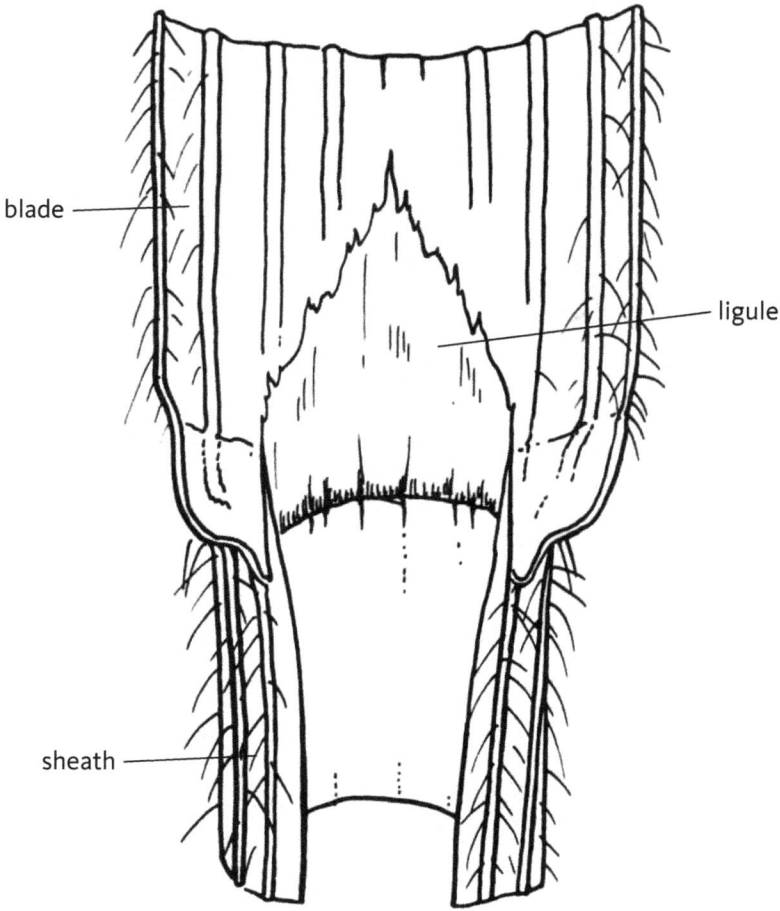

KEY CHARACTERS Erect annual. Sheath closed. Blades hairy. Ligule hyaline, large 2–5 mm, obtuse-lacerate. Awns, 5–10 mm.

VERNATION Curled. **BLADES** Flat, wide, long; hairy ventrally and dorsally; veins each side of midrib, 2–3; ribs prominent ventrally and dorsally; margin hairy; midrib prominent dorsally; 4–6 mm wide, 10–20 cm long. **AURICLE** None. **LIGULE** Membranous, hyaline, large 2–5 mm, obtuse-lacerate. **COLLAR** Glabrous, divided. **SHEATH** Closed, hairy. **NODE** Glabrous. **INTERNODE** Glabrous, veined. **ROOTS** Fibrous, annual. **CULM** Round, not branched.

HABITAT Dry, open places and disturbed areas.

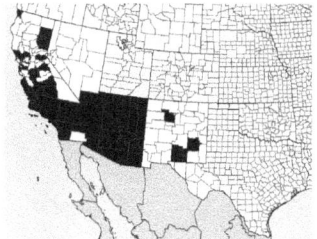

Bromus ciliatus L.

FRINGED BROME

Bromus ciliatus L.
FRINGED BROME

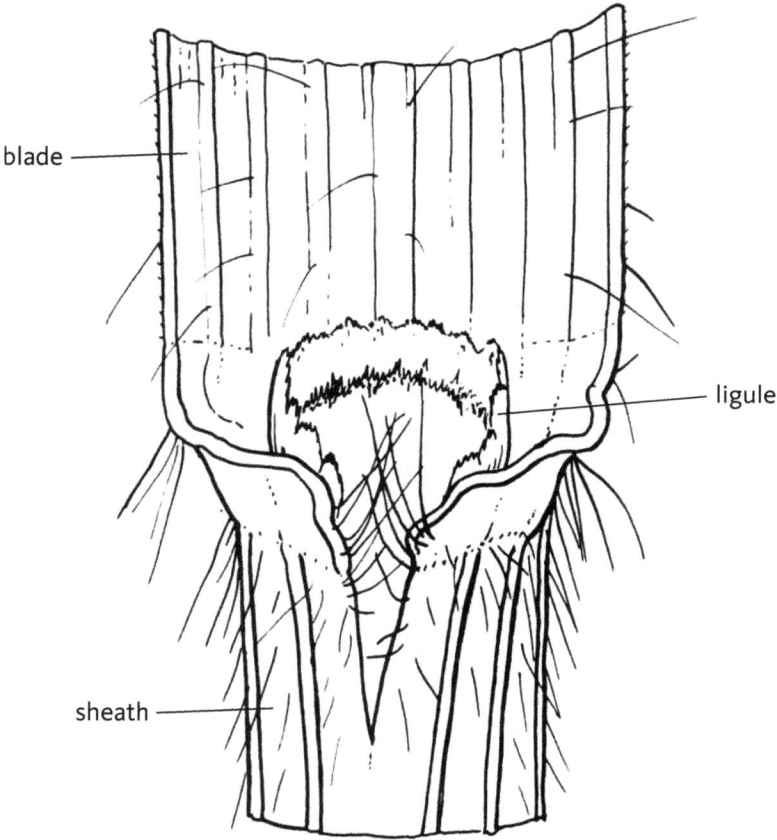

blade

ligule

sheath

KEY CHARACTERS Small, erect bunchgrass, panicle drooping. Sheath closed lower 2/3. Ligule truncate-entire, 1/2–1 mm. Conspicuous soft hairs over plant. Midrib prominent dorsally and ventrally.

VERNATION Curled. **BLADES** Flat, narrow, long, drooping; hairy; veins each side of midrib, 2–3; ribs prominent ventrally and dorsally; midrib prominent dorsally and ventrally; 2–8mm wide, 15–30cm long. **AURICLE** None. **LIGULE** Membranous, small 2/3–1 mm, truncate-notched. **COLLAR** Hairy on margin. **SHEATH** Hairy veined, usually closed in lower 2/3. **NODE** Glabrous. **INTERNODE** Glabrous. **ROOTS** Fibrous. **CULM** Round, not branched.

HABITAT Moist meadows, thickets and woods, streambanks.

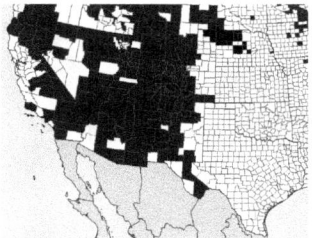

22 *Bromus inermis* Leyss.
SMOOTH BROME

Bromus inermis Leyss.

SMOOTH BROME

blade

ligule

sheath

....

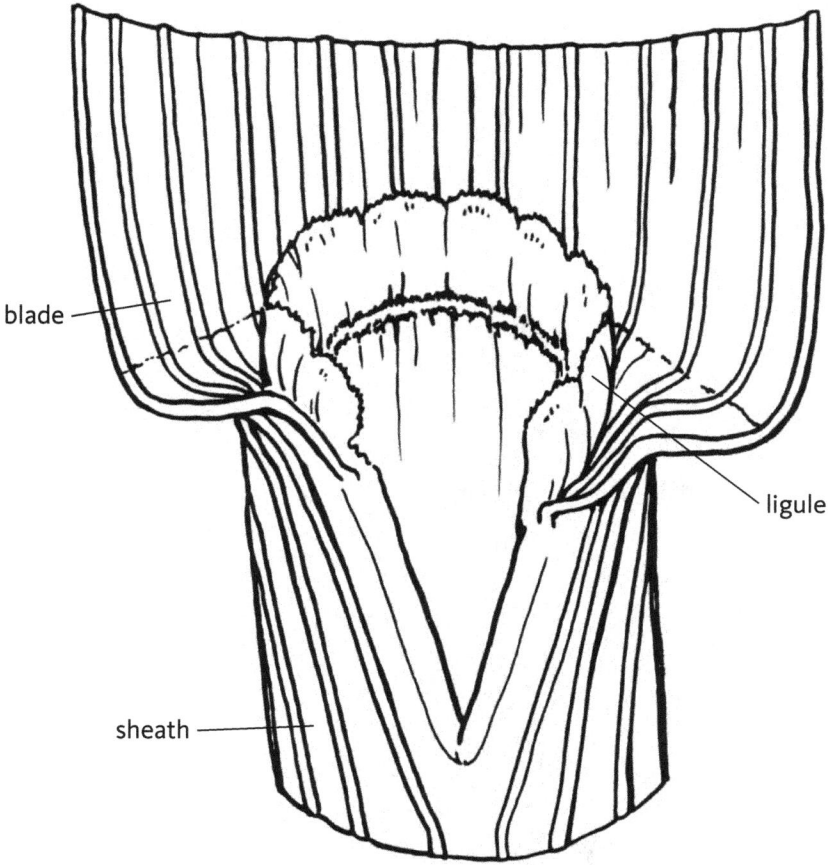

KEY CHARACTERS Introduced, large, erect, sod-former. Sheath closed. The letter 'W' or 'M' on blades. Blades large, flat, long ribbed.

....

VERNATION Curled. **BLADES** Flat, wide, drooping, usually a 'W' on blade; rough ventrally and dorsally; veins each side of midrib, 5-6; rib prominent ventrally and dorsally; midrib prominent dorsally; 4-5 mm wide, 10-20 cm long. **AURICLE** None.

LIGULE Membranous, small, 1/2 mm. **COLLAR** Glabrous, divided. **SHEATH** Closed, sometimes hairy, veined, round, margin papery. **NODE** Occasionally pubescent. **INTERNODE** Glabrous, veined. **ROOTS** Short rhizomes. **CULM** Round, not branched, large.

....

HABITAT Pastures, fields, sometimes used in restoration projects.

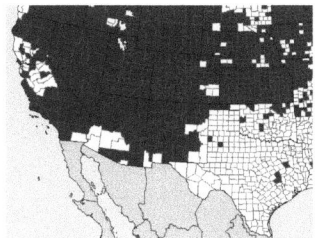

RED BROME

Bromus rubens L.
RED BROME

blade

ligule

sheath

···

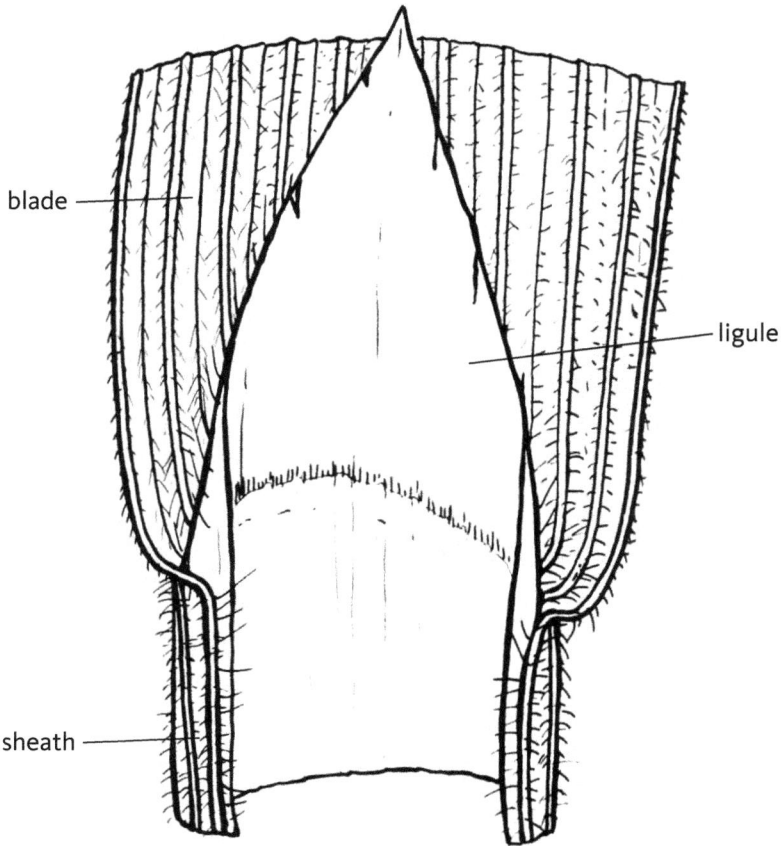

KEY CHARACTERS Introduced, erect to semi-decumbent, winter annual. Blades short, narrow, flat. Midrib prominent ventrally. Ligule acute-lacerate. Awns 1/2–1 cm. Head at maturity reddish color.

···

VERNATION Curled. **BLADES** Flat, pointed; hairy dorsally and ventrally; veins each side midrib, 2–3; ribs prominent ventrally and dorsally; blade margin smooth; midrib prominent, 1–2 mm wide, 2–6 cm. long. **AURICLE** None. **LIGULE** Membranous, 1/2–3 mm long, acute-lacerate. **COLLAR** Smooth. **SHEATH** Pubescent, margin papery, round, usually closed. **NODE** Glabrous. **INTERNODE** Occasionally puberulent near nodes. **ROOTS** Fibrous, weak. **CULM** Not branched, round.

···

HABITAT Disturbed and waste places, fields, rocky slopes.

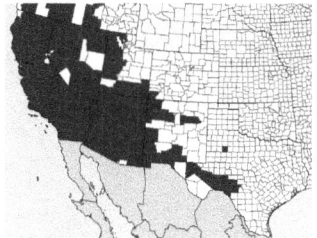

24 *Bromus tectorum* L.
DOWNY CHESS, CHEATCRASS BROME

DOWNY CHESS, CHEATCRASS BROME

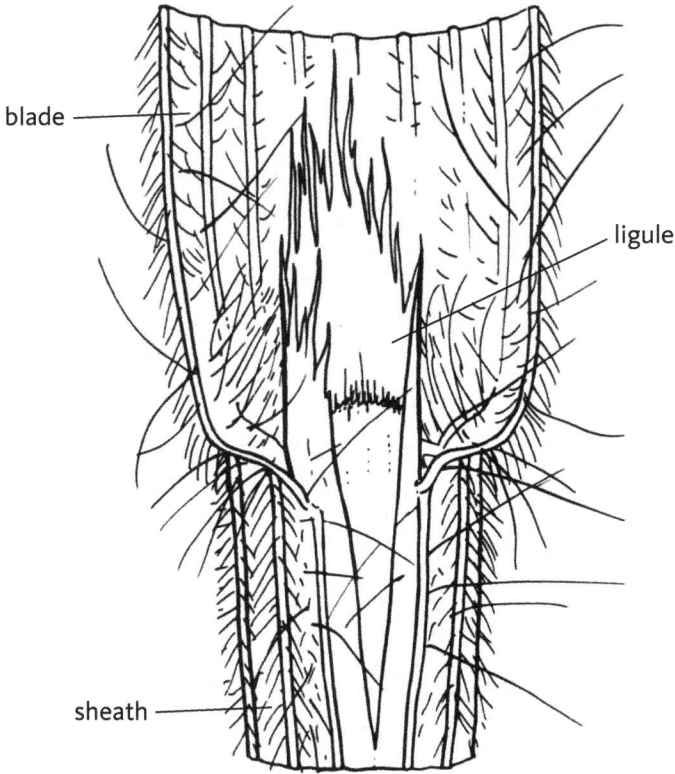

24

KEY CHARACTERS Introduced, semi-erect, small bunchgrass, winter annual. Heads open, drooping. Awn, 1–2 cm. Fire hazard. Sheath closed. Ligule very thin, obtuse-lacerate. Woolly over most of plant, soft to touch. In northern climates grows in spring and fall. Excellent for lambing.

VERNATION Curled. BLADES Flat or twisted, drooping, blunt pointed; hairy ventrally and dorsally, soft; veins each side of midrib, 2–3; ribs indistinct; margin hairy; midrib prominent ventrally; 2–4 mm wide, 5–10 cm long. AURICLE None. LIGULE Membranous, collar-like, small, 1–2 mm, obtuse-lacerate, paper-like. COLLAR Smooth, divided. SHEATH Woolly, round, closed. NODE Glabrous, dark in color. INTERNODE Glabrous. ROOTS Shallow, fibrous. CULM Round, not branched.

HABITAT Disturbed places, such as overgrazed rangelands, fields, sand dunes, roadsides, and waste areas. Common in sagebrush habitats following fire.

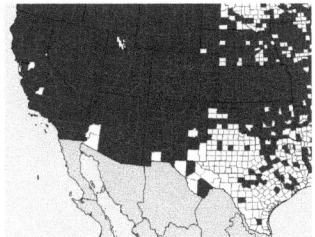

25 *Chloris virgata* Sw.
FEATHER FINGERGRASS

Chloris virgata sw.
FEATHER FINGERGRASS

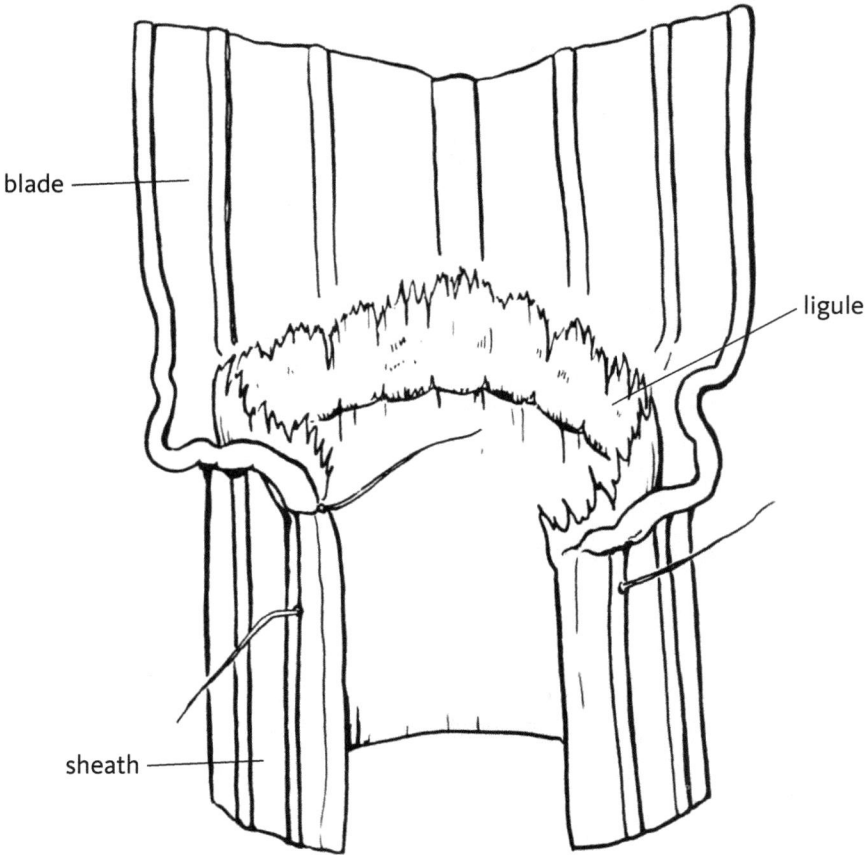

blade

ligule

sheath

....

KEY CHARACTERS Small, annual, semi-erect bunchgrass. Blades flat, wide, occasionally with glandular hairs dorsally. Ligule 1 mm, truncate-ciliate. Blade midrib prominent ventrally.

....

VERNATION Curled. **BLADES** Flat, wide, long; occasionally glandular hairs dorsally near base, 2–3 mm; veins each side of midrib, 3; ribs numerous, not prominent; margin often straw colored; midrib prominent ventrally; 3–6 mm wide, 8–15 cm long. **AURICLE** None. **LIGULE** Membranous, small, 1 mm, truncate-ciliate. **COLLAR** Glabrous. **SHEATH** Glabrous, hyaline margin. **NODE** Glabrous. **INTERNODE** Glabrous. **ROOTS** Fibrous, annual. **CULM** Elliptical, branched at base.

....

HABITAT Widespread species of tropical to temerate areas; in our region, a common weed in alfalfa fields.

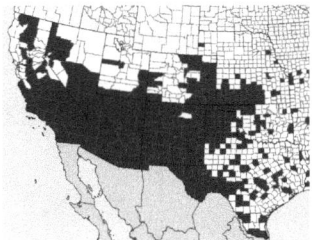

79

BERMUDAGRASS

Cynodon dactylon (L.) Pers.
BERMUDAGRASS

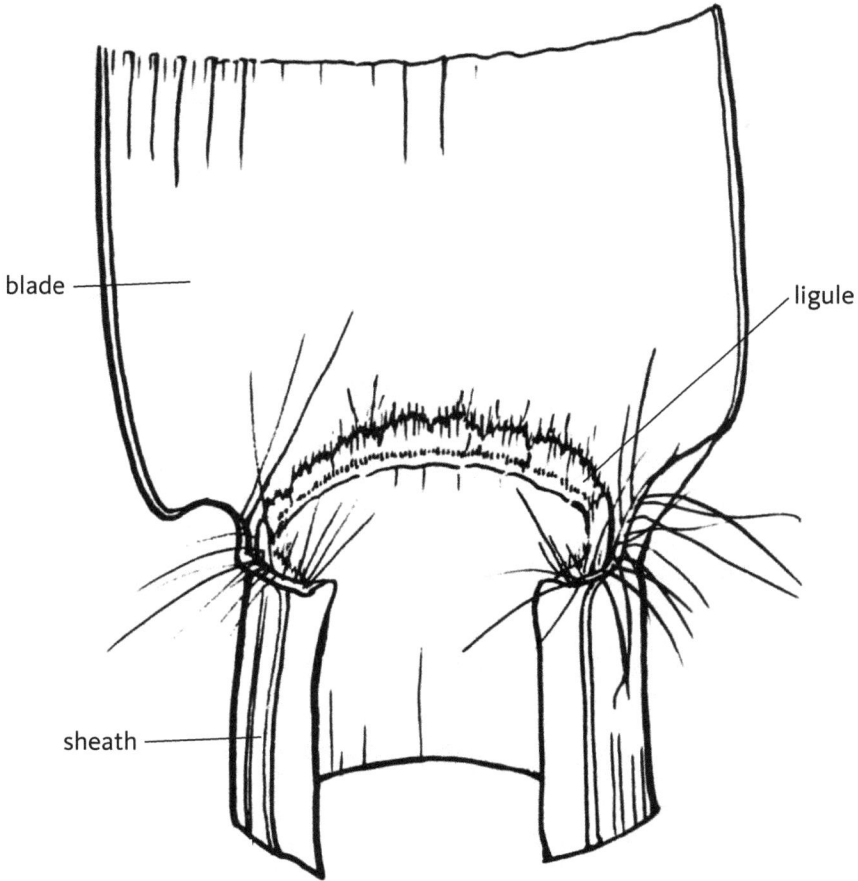

blade

ligule

sheath

KEY CHARACTERS Sod-forming. Culms stoloniferous, branched. Ligule hairy. Blade flat, short, wide. Growth decumbent except flower stalks.

VERNATION Curled. **BLADES** Flat, wide, short, usually glabrous; veins each side of midrib, 2; ribs not prominent; margin white, midrib prominentdorsally; 1-4 mm wide, 2-6 cm long. **AURICLE** None. **LIGULE** Hairy, 1 mm, with scattered long hairs.

COLLAR Usually glabrous, occasionally with long hairs. **SHEATH** Usually glabrous, hyaline margin, often with tufts of hair at the summit. **NODE** Glabrous. **INTERNODE** Glabrous. **ROOTS** Rhizomatous. **CULM** Elliptical, branched, stoloniferous.

HABITAT Common weed of disturbed habitats; used as a lawn grass.

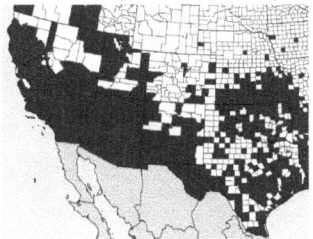

Dasyochloa pulchella (Kunth) Willd. ex Rydb.
FLUFFGRASS

Dasyochloa pulchella (Kunth) Willd. ex Rydb.
FLUFFGRASS

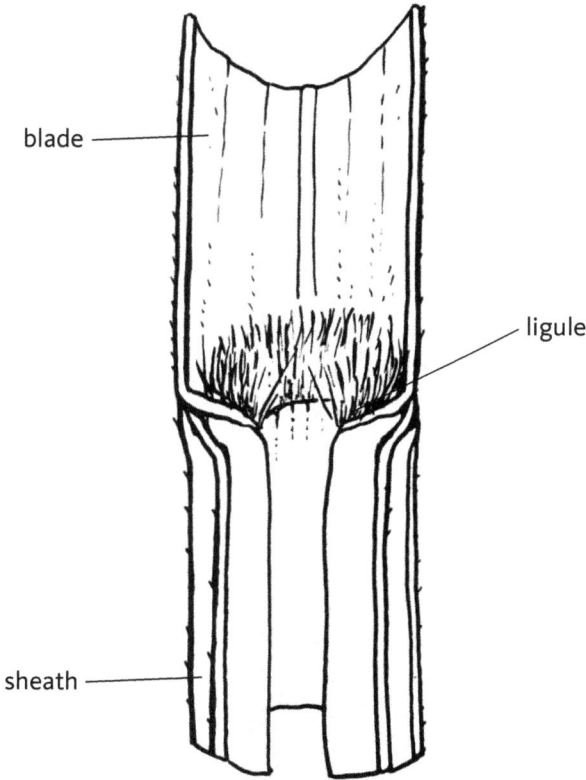

SYNONYM *Tridens pulchellus* (Kunth) A.S. Hitchc.

KEY CHARACTERS Small, erect, bunchgrass. Blades ribbed, rolled, small and short. Sheath margin conspicuously paper-like, long ciliate. Flowers clustered, somewhat exceeded by the leaves, cobwebby. Rooting at nodes.

VERNATION Folded. **BLADES** Rolled, erect, very small, narrow, pointed; rough ventrally; ribs distinct, veins indistinct; margin toothed; width 1/4 mm, length 2–5 cm. **AURICLE** None. **LIGULE** Hairy, small, 1/2 mm. **COLLAR** Smooth, small, entire. **SHEATH** Margin paper-like, the lower long ciliate. **NODE** Glabrous, **ROOTS** at nodes. **INTERNODE** Glabrous to pubescent. **ROOTS** Fibrous. **CULM** Elliptical, terminal **NODE** branched with cluster of flowers often cobwebby.

HABITAT Rocky soils of arid regions; a common grass in areas dominated by creosote-bush.

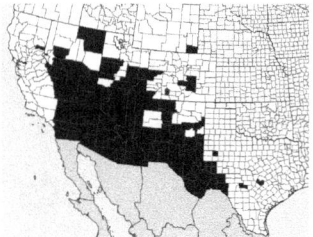

28 *Digitaria californica* (Benth.) Henr.
ARIZONA COTTONTOP

Digitaria californica (Benth.) Henr.

ARIZONA COTTONTOP

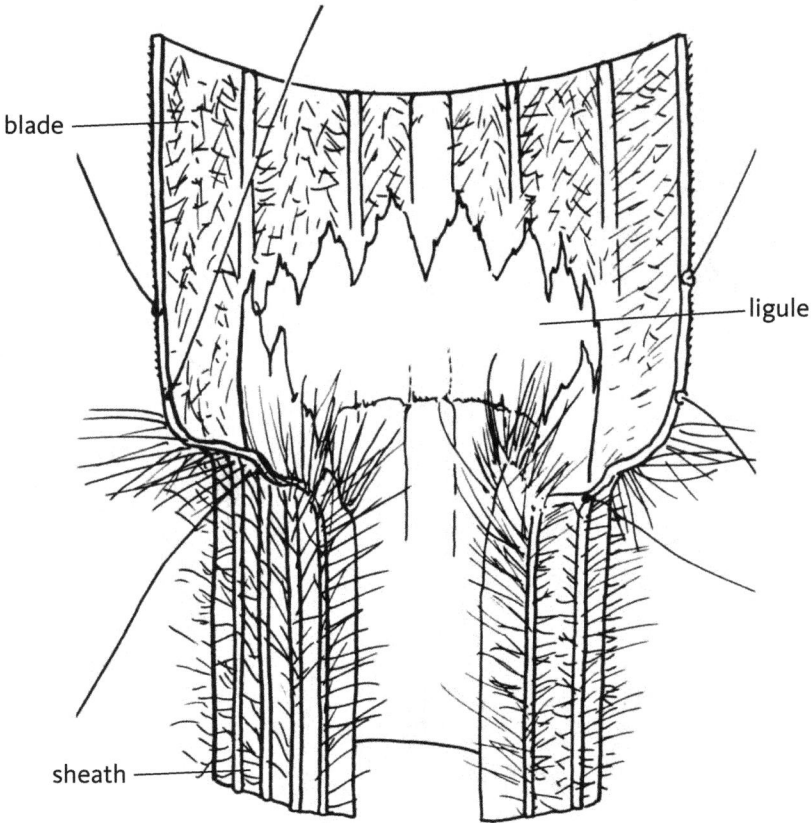

SYNONYM *Trichachne californica* (Benth.) Chase

KEY CHARACTERS Erect bunchgrass. Blades flat, long; blade margin white. Culms branched. Ligule 2–3 mm, obtuse-lacerate.

VERNATION Curled. BLADES Flat, drooping, narrow, pointed; glabrous or pubescent ventrally and dorsally, occasional glandular hairs dorsally; veins each side of midrib 3; ribs prominent ventrally; margin toothed, white; midrib prominent; width 3–4 mm, length 8–12 cm. AURICLE None. LIGULE Membranous, 2–3 mm, obtuse-lacerate. COLLAR Hairy ventrally, entire. SHEATH Glabrous or pubescent, elliptical, margin hairy. NODE Pubescent, occasionally glabrate. INTERNODE Glabrous. ROOTS Fibrous. CULM Elliptical, branched.

HABITAT Dry plains.

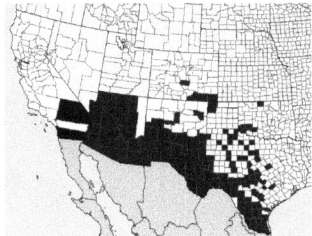

Distichlis spicata (L.) Greene
DESERT SALTGRASS

♀

♂

Distichlis spicata (L.) Greene

DESERT SALTGRASS

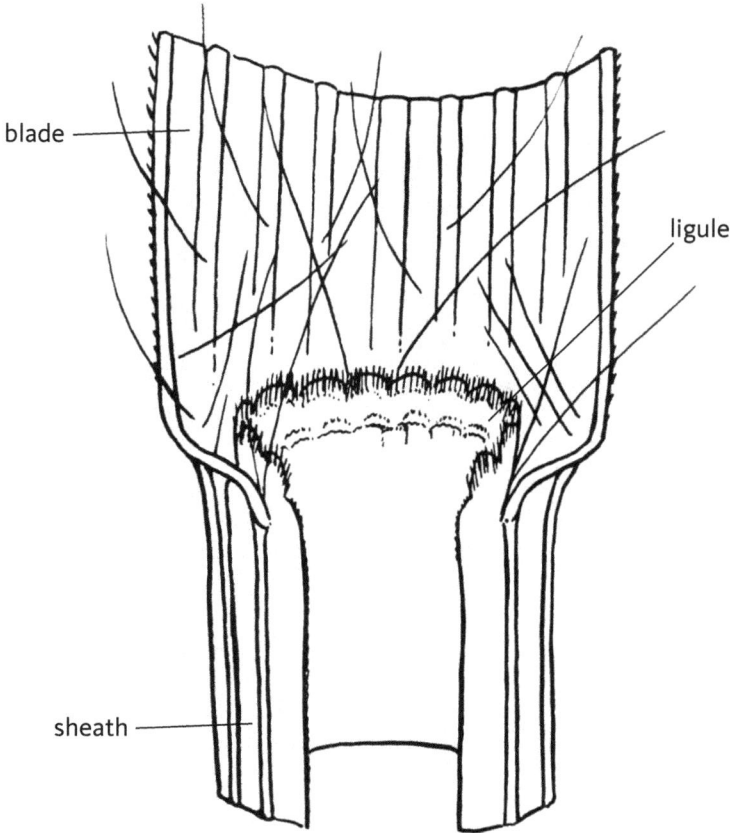

blade

ligule

sheath

....

SYNONYM *Distichlis stricta* (Torr.) Rydb.

....

KEY CHARACTERS Erect, semi-sodgrass. Rhizomatous. Blades usually rolled at tip, short. Lower culms and rhizomes conspicuously shiny.

....

VERNATION Clasped. **BLADES** Flat to rolled, erect, narrow, pointed; hairy ventrally, frequently with salt deposits; veinseach sideof midrib,3; ribs prominentdorsally; margin toothed; midrib prominent ventrally; 2-3 mm wide at base, 1-6 cm long.

AURICLE None. **LIGULE** Hairy, small, 1/2-1 mm. **COLLAR** Occasionally hairy margin, hairs 1-2 mm. **SHEATH** Smooth, round, margin papery. **NODE** Glabrous. **INTERNODE** Glabrous. **ROOTS** Rhizomatous. **CULM** Round, not branched.

....

HABITAT Saline soils, sometimnes where seasonally wet.

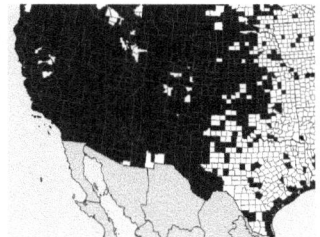

30 *Elionurus barbiculmis* Hack.
WOOLSPIKE

WOOLSPIKE

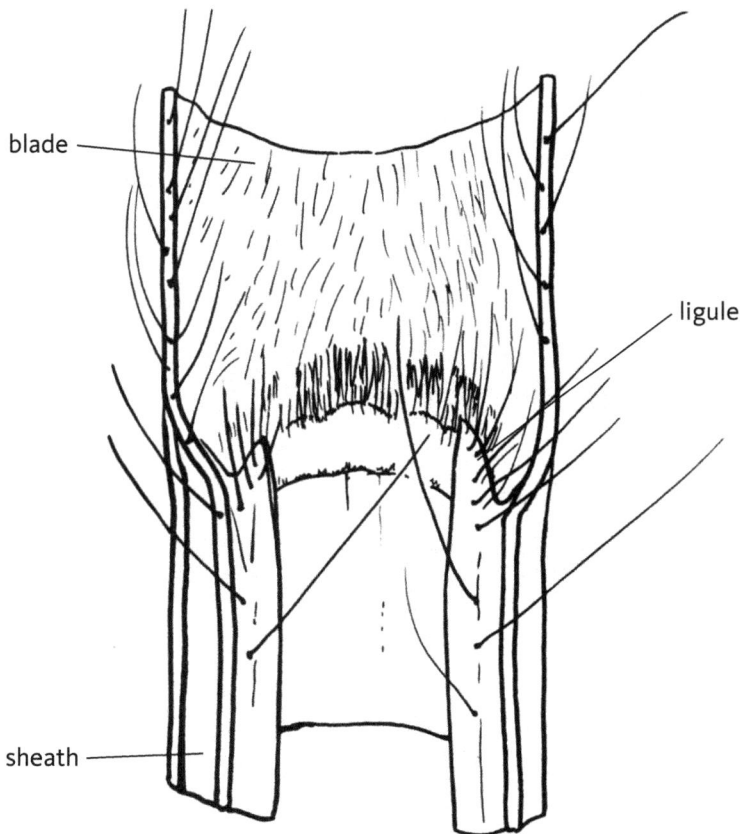

SYNONYM *Elyonurus barbiculmis* Hack.

KEY CHARACTERS Erect, medium bunchgrass. Ligule hairy, front pilose, 2–5 mm. Stems pubescent below node. Blades narrow, long rolled, hairy ventrally. Culms branched; heads silvery, densely pilose.

VERNATION Folded. **BLADES** Rolled, narrow, long, hairy, ventrally, pilose, 3–5 mm; ribs prominent dorsally; midrib not prominent; 1 mm wide, 15–30 cm long. **AURICLE** None. **LIGULE** Hairy, small, 1 mm, front pilose, 1–5 mm. **COLLAR** Glabrous, divided. **SHEATH** Glabrous, veined. **NODE** Glabrous. **INTERNODE** Pubescent to villous below node. **ROOTS** Fibrous. **CULM** Round, branched.

HABITAT Mesas, rocky slopes, hills, canyons.

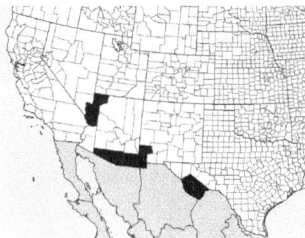

Elymus elymoides (Raf.) Swezey

BOTTLEBRUSH SQUIRRELTAIL

Elymus elymoides (Raf.) Swezey
BOTTLEBRUSH SQUIRRELTAIL

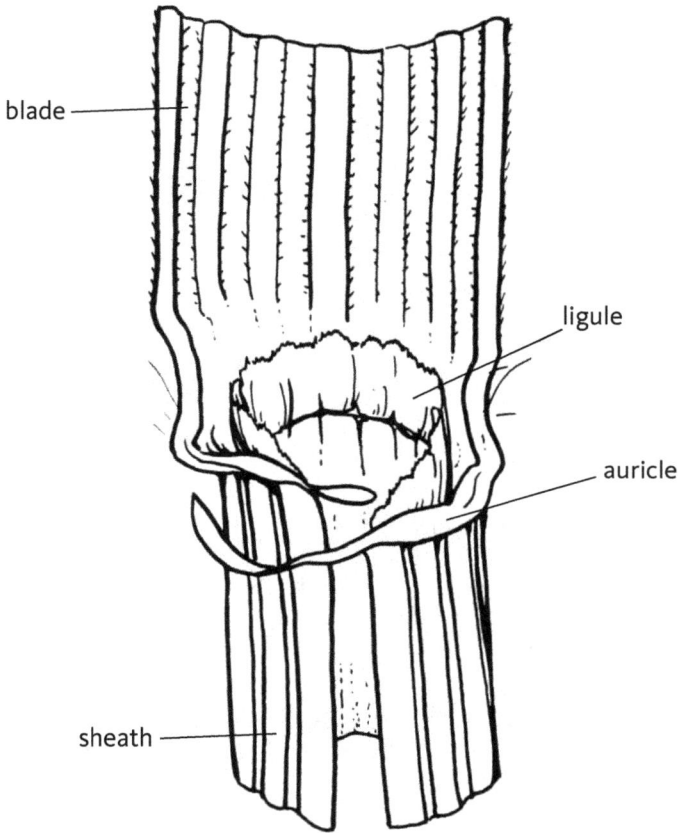

SYNONYM *Sitanion hystrix* (Nutt.) J.G. Sm.

KEY CHARACTERS Small, semi-erect bunchgrass. Auricle present. Ribs prominent dorsally.

VERNATION Curled. BLADES Usually flat, twisted, drooping, narrow, pointed; rough, pubescent to glabrous; veins each side of midrib, 2–4; ribs prominent dorsally; 2–3 mm wide, 5–20 cm long. AURICLE Small, 2–3 mm. LIGULE Membranous, small, 1/4 mm, truncate-entire. COLLAR Glabrous, divided. SHEATH Pubescent or glabrous, ribbed, round frequently pinkish. NODE Glabrous. INTERNODE Glabrous. ROOTS Fibrous. CULM Round, not branched.

HABITAT Dry, often rocky, open woods, thickets, grasslands, and disturbed areas; common in overgrazed pinyon-juniper woodlands.

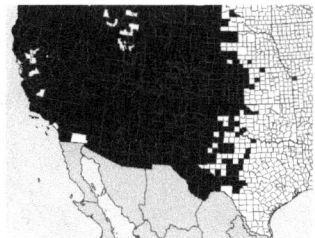

PLAINS LOVEGRASS

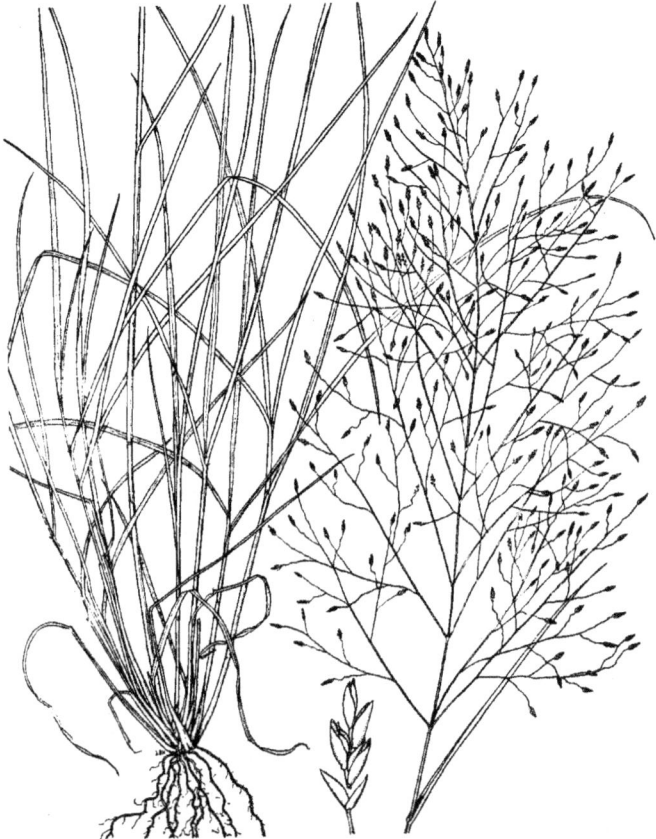

Eragrostis intermedia A.S. Hitchc.
PLAINS LOVEGRASS

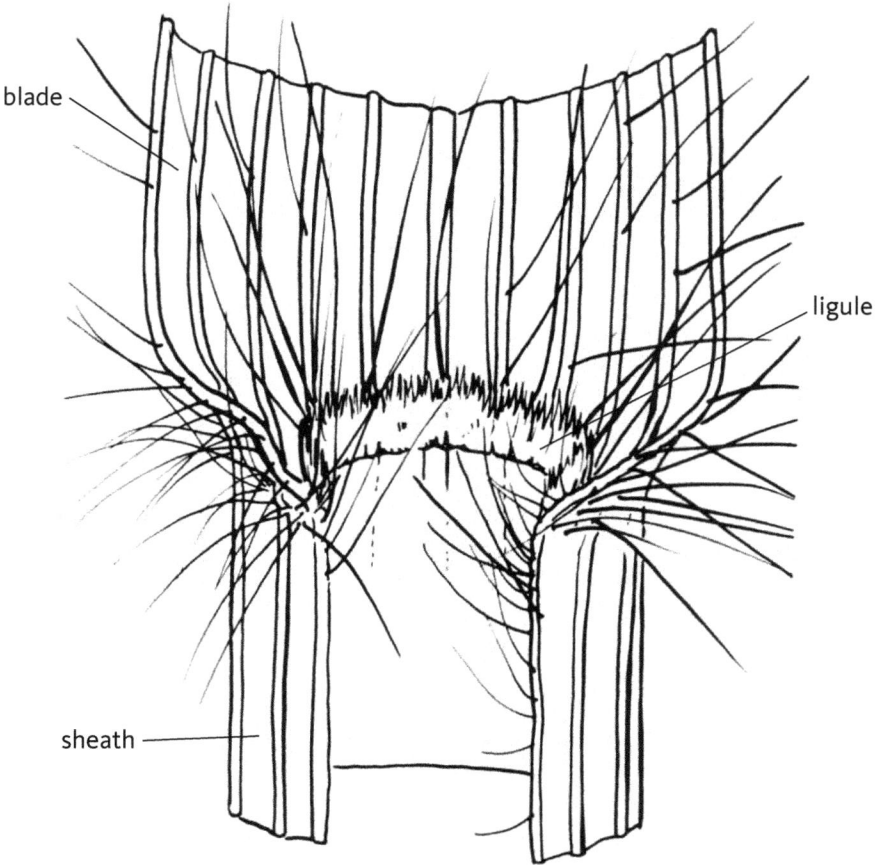

KEY CHARACTERS Erect, medium bunchgrass. Blades flat, long narrow. Conspicuous ribs on blade and sheath. Ligule membranous with long hairs. Collar bearded. Culm elliptical.

VERNATION Curled. **BLADES** Flat, long, narrow, scattered hairs; veins each side of midrib, 3-4; ribs prominent ventrally and dorsally; margin smooth; midrib prominent dorsally; 2 mm wide, 10-30 cm long. **AURICLE** None. **LIGULE** Membranous, small, 1/2 mm, truncate-ciliate, with long conspicuous hairs, 3-5 mm. **COLLAR** Hairy on margin. **SHEATH** Glabrous, veined. **NODE** Glabrous. **INTERNODE** Glabrous. **ROOTS** Fibrous. **CULM** Elliptical, not branched.

HABITAT Disturbed places.

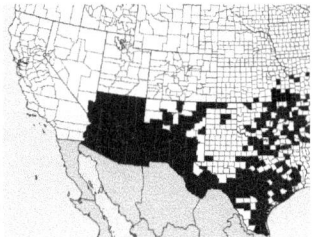

Eragrostis lehmanniana Nees.

LEHMANN LOVEGRASS

Eragrostis lehmanniana Nees.
LEHMANN LOVEGRASS

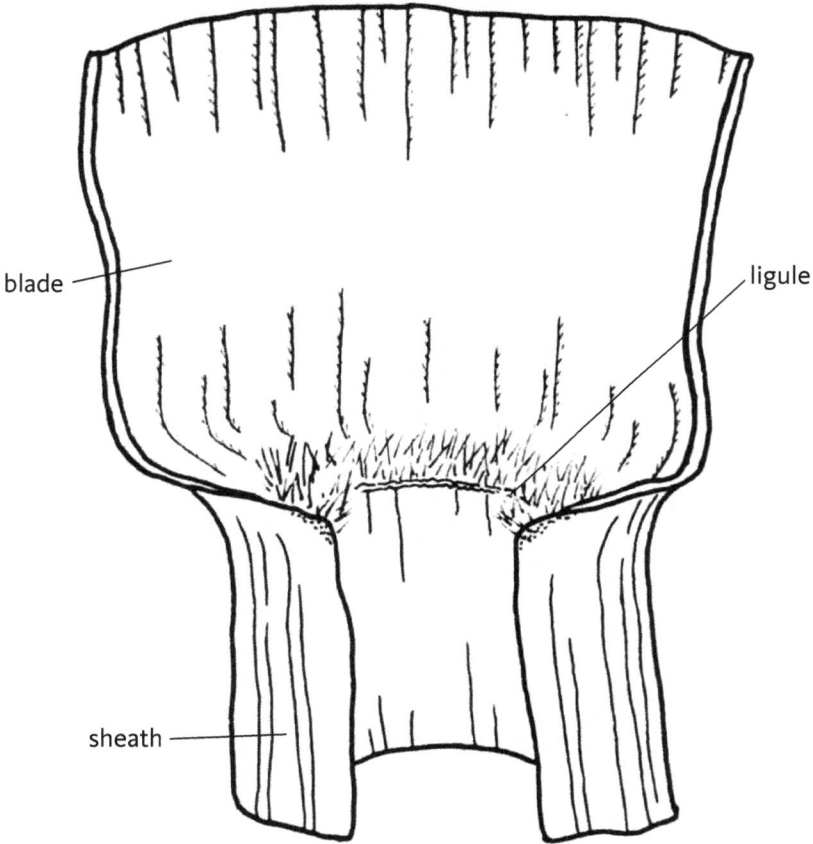

blade

ligule

sheath

....

KEY CHARACTERS Large, erect, bunchgrass. Blades flat, long; blade ribs prominent. Collar margin hairy, 1–3 mm. Ligule truncate-ciliate, 1/2–1 mm. Culm round.

....

VERNATION Curled. **BLADES** Flat, drooping, wide, long, pointed; glabrous, soft; veins each side of midrib, 3; ribs prominent ventrally and dorsally; margin glabrous; midrib not prominent, 3–4 mm wide, 12–15 cm long. **AURICLE** None. **LIGULE** Hairy or membranous, truncate-ciliate, 1/2–1 mm. **COLLAR** Hairy margin, 2–3 mm. **SHEATH** Round, veined, frequently bearded at base. **NODE** Occasionally pubescent. **INTERNODE** Glabrous. **ROOTS** Fibrous. **CULM** Round, occasionally branched.

....

HABITAT Sandy flats, roadsides, calcareous slopes, disturbed areas. Introduced from Africa for erosion control.

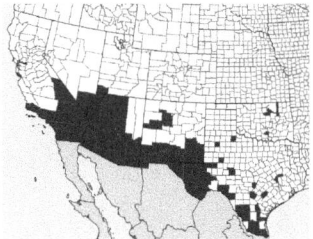

Festuca arizonica Vasey

ARIZONA FESCUE

blade

ligule

sheath

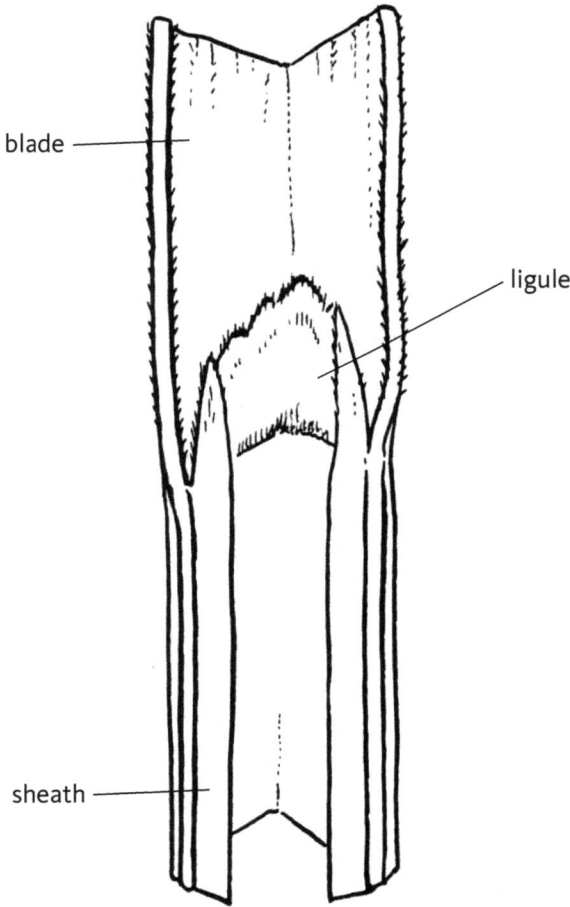

....
KEY CHARACTERS Large, erect, bunchgrass. Blades long drooping, rolled, thread-like. Ligule very small and difficult to observe.
....
VERNATION Folded. **BLADES** Rolled, drooping, very small; glabrous; ribs indistinct; margin toothed; midrib indistinct; 1/4–1/2 mm wide, 10–30 cm long. **AURICLE** None.
LIGULE Membranous; very small, 1/4–1/2 mm; acute-entire. **COLLAR** Indistinct. **SHEATH** Elliptical, pinkish above roots. **NODE** Glabrous. **INTERNODE** Glabrous. **ROOTS** Fibrous. **CULM** Round, not branched.
....
HABITAT Dry meadows, montane forest openings; soils gravelly or rocky.

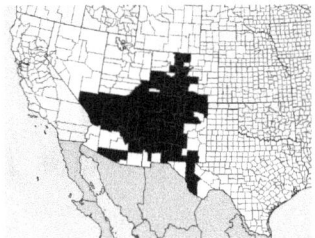

35 *Hesperostipa neomexicana* (Thurb. ex Coult.) Barkworth
NEW MEXICAN FEATHERGRASS

NEW MEXICAN FEATHERGRASS

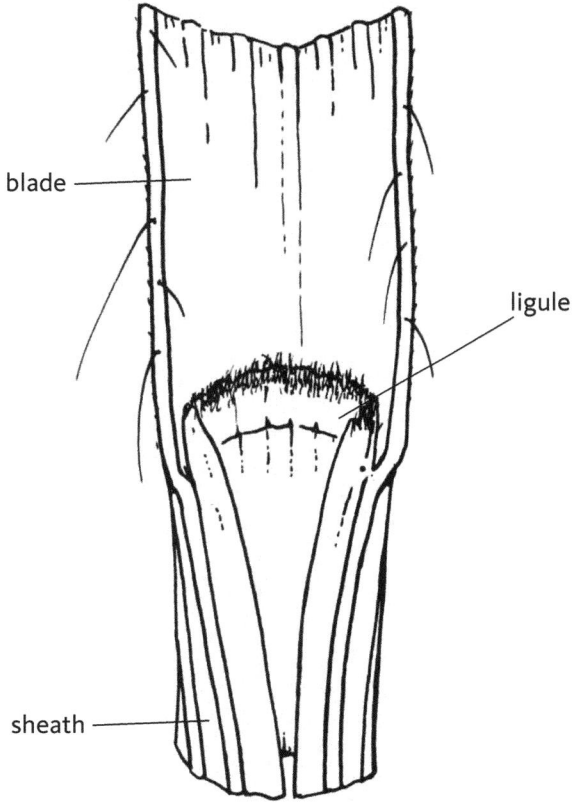

SYNONYM *Stipa neomexicana* (Thurb. ex Coult.) Scribn.

KEY CHARACTERS Erect, medium-sized bunchgrass. Blades long, needle-like. Ligule very small, 1/4 mm, truncate-ciliate. Awns long, twisted with long hairs toward tip.

VERNATION Folded (also reported as curled). **BLADES** Rolled, drooping, narrow, pointed; glabrous; veins indistinct; midrib not prominent; width 1/2–1 mm, length 25–35 cm. **AURICLE** None. **LIGULE** Membranous, very small, 1/4 mm, truncate-ciliate. **COLLAR** Glabrous, margin appearing hairy, divided, indistinct. **SHEATH** Elliptical, veined. **NODE** Glabrous. **INTERNODE** Glabrous. **ROOTS** Fibrous. **CULM** Elliptical, not branched.

HABITAT Grasslands, oak and pinyon pine woodlands; soils well-drained and often rocky.

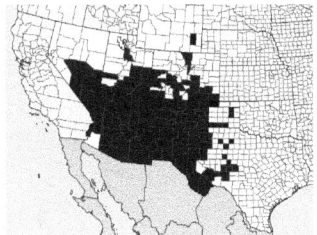

Heteropogon contortus (L.) Beauv. ex Roemer & J.A. Schultes

TANGLEHEAD

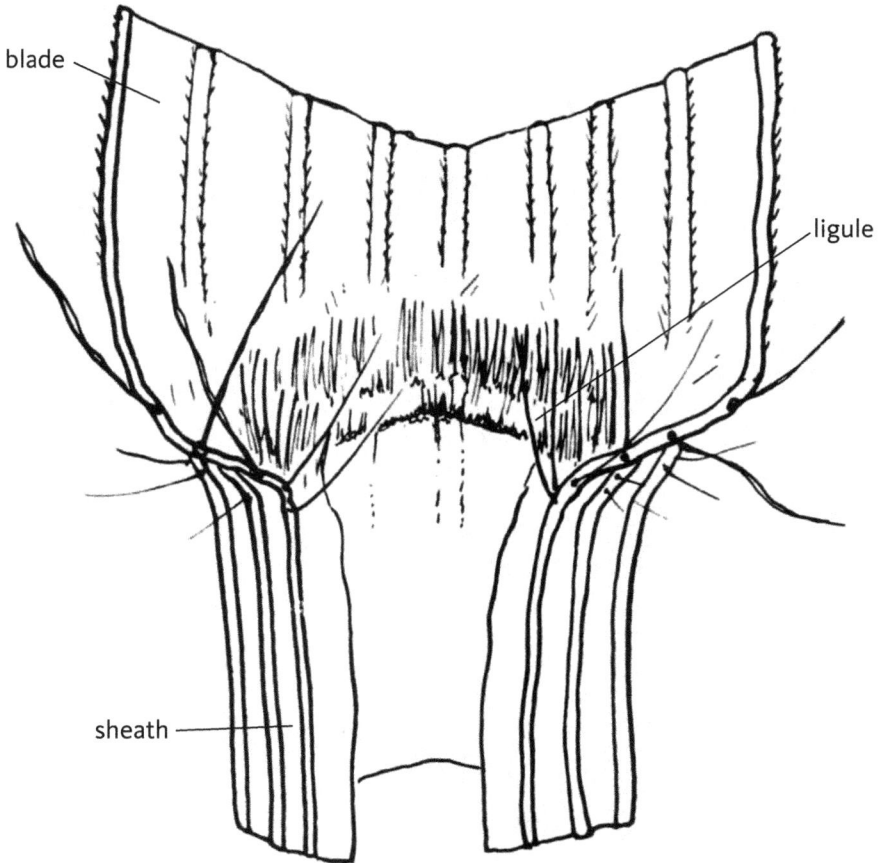

KEY CHARACTERS Erect, small bunchgrass. Blades flat, wide, occasionally folded at base; blade margin and collar with glandular hairs. Culm flat. Ligule hairy, small. Midrib prominent ventrally. Awns long, hairy, forming a tangled mass.

VERNATION Folded. **BLADES** Flat, wide; hairy occasionally ventrally; veins each side of midrib, 3-4; ribs not prominent; margin glandular at base, white; midrib prominent ventrally; 6-8 mm wide, 6-15 cm long. **AURICLE** None. **LIGULE** Hairy, lower 1/2 membranous, overall length 1 mm. **COLLAR** Margin hairy, glandular, 2-4 mm. **SHEATH** Flat, glabrous, veined. **NODE** Glabrous. **INTERNODE** Glabrous. **ROOTS** Fibrous. **CULM** Flat, not branched.

HABITAT Rocky hills and canyons, disturbed places.

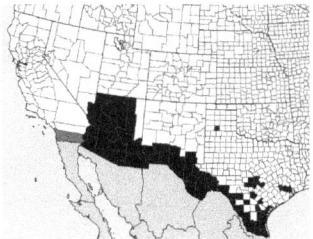

37 *Hilaria belangeri* (Steud.) Nash.
CURLY-MESQUITE

Hilaria belangeri (Steud.) Nash.

CURLY-MESQUITE

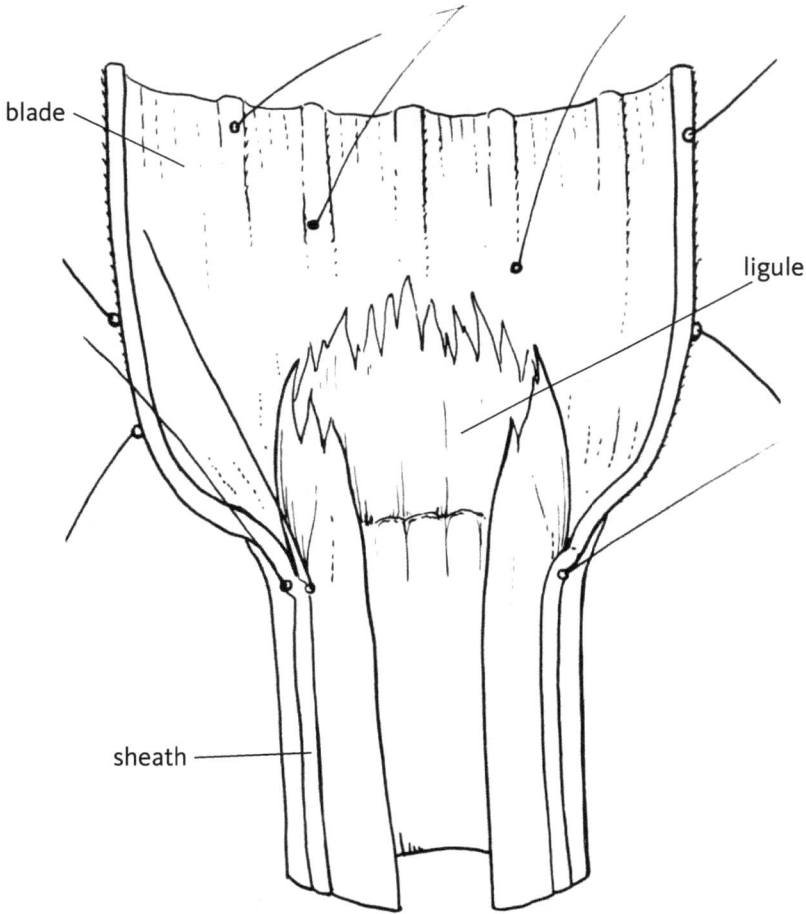

blade

ligule

sheath

....

KEY CHARACTERS Loose sod-forming grass. Culms rooting at lower nodes. Nodes hairy. Blades with glandular hairs, 2–4 mm. Ligule 2 mm, truncate-lacerate.

....

VERNATION Curled. **BLADES** Flat, curled, narrow, pointed, long; hairy, glandular; veins on each side of midrib, 2; ribs not prominent ventrally; 2 mm wide, 5–20 cm long. **AURICLE** None. **LIGULE** Membranous, large, 2 mm; truncate-lacerate. **COLLAR** Glabrous. **SHEATH** Glabrous, paper margin. **NODE** Hairy. **INTERNODE** Glabrous. **ROOTS** Fibrous. **CULM** Round, lower nodes grow **ROOTS** and culms.

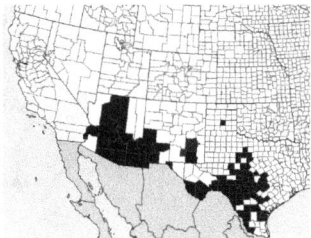

....

HABITAT plains and mesas; a dominant grass on Texas shortgrass prairies.

38 *Hilaria jamesii* (Torr.) Benth.
GALLETA

Hilaria jamesii (Torr.) Benth.

GALLETA

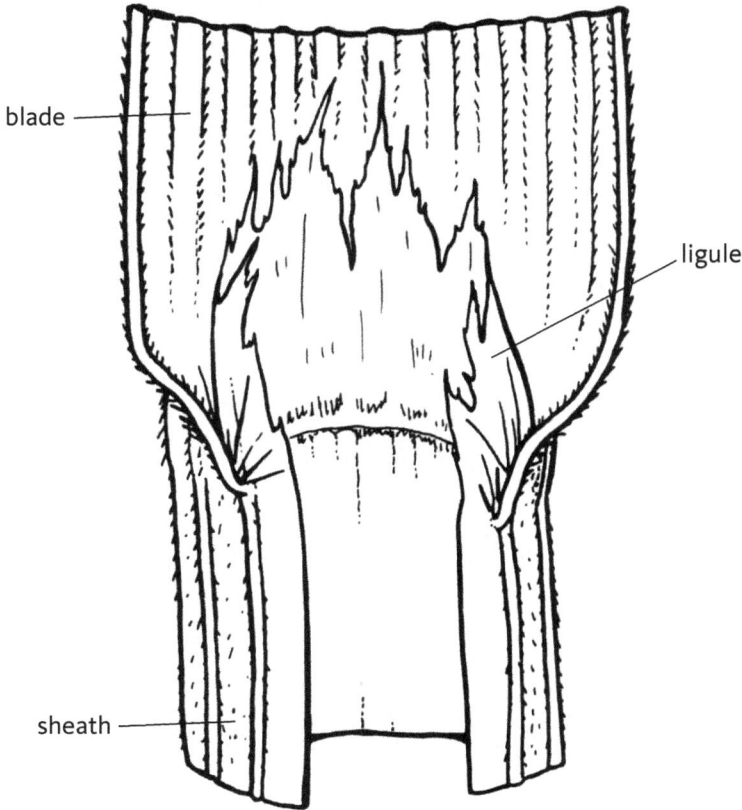

blade

ligule

sheath

SYNONYM *Pleuraphis jamesii* Torr.

KEY CHARACTERS Erect, large bunchgrass. Blades flat at base, upper rolled. Rhizomatous. Nodes pubescent. Ligule 1-3 mm, truncate-lacerate. Culm base frequently semi-decumbent. Rachis zigzag.

VERNATION Curled. **BLADES** Flat base, upper 2/3 often rolled, pointed; rough ventral and dorsal; veins each side of midrib, 3-4; ribs prominent dorsally; margin toothed; midrib not prominent; 3-5 mm wide at base, 4-12 cm long. **AURICLE** None. **LIGULE** Membranous, 1-3 mm, truncate-lacerate. **COLLAR** Usually glabrous. **SHEATH** Elliptical, veined. **NODE** Pubescent to villous. **INTERNODE** Often pubescent below node. **ROOTS** Rhizomatous. **CULM** Round, not branched.

HABITAT Deserts, canyons, dry plains.

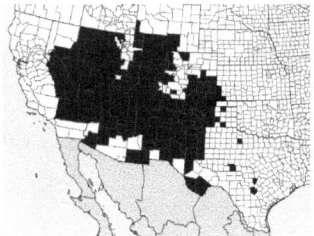

105

Hilaria mutica (Buckl.) Benth.

TOBOSA

blade

ligule

sheath

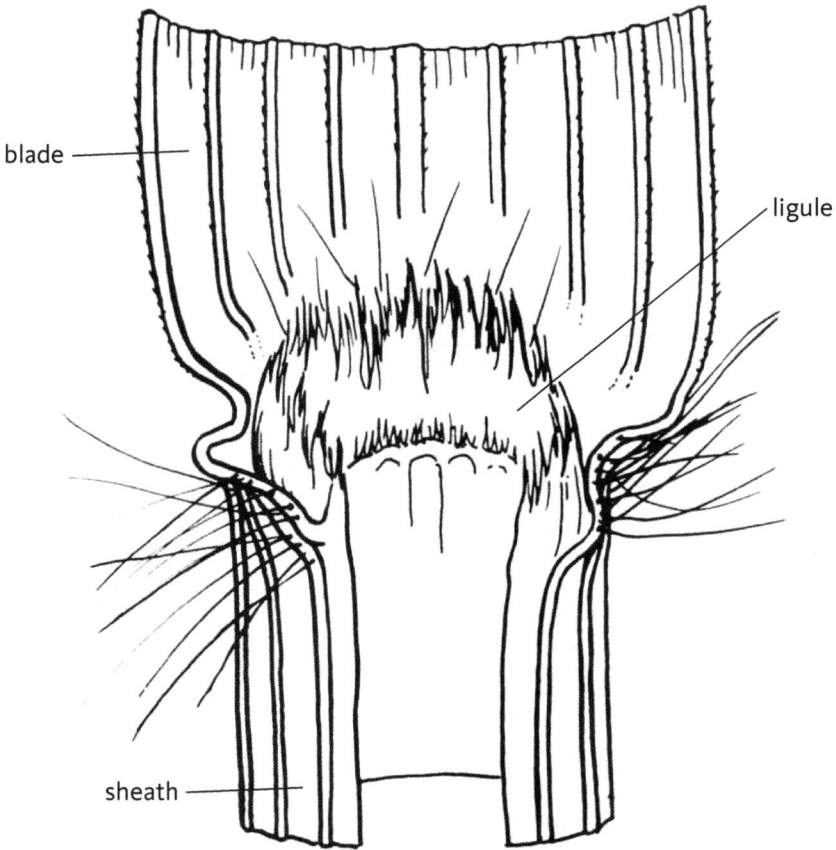

SYNONYM *Pleuraphis mutica* Buckl.

KEY CHARACTERS Large bunchgrass, semi-erect. Blade usually rolled. Rhizomatous. Culms branched. Ligule truncate-ciliate.

VERNATION Curled. **BLADES** Flat to rolled; occasionally hairy ventrally; veins each side of midrib, 2–3; ribs prominent; midrib prominent ventrally; 2–4 mm wide, 5–10 cm long. **AURICLE** None. **LIGULE** Membranous, truncate-ciliate, 1–2 mm. **COLLAR** Margin hairy. **SHEATH** Glabrous, veined. **NODE** Glabrous to pubescent. **INTERNODE** Glabrous. **ROOTS** Rhizomatous. **CULM** Round, Branched.

HABITAT Upland flats and desert valleys, usuallywhere subject to occasional flooding.

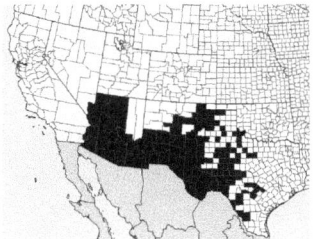

40 *Hilaria rigida* (Thurb.) Benth. ex Scribn.
BIG GALLETA

Hilaria rigida (Thurb.) Benth. ex Scribn.
BIG GALLETA

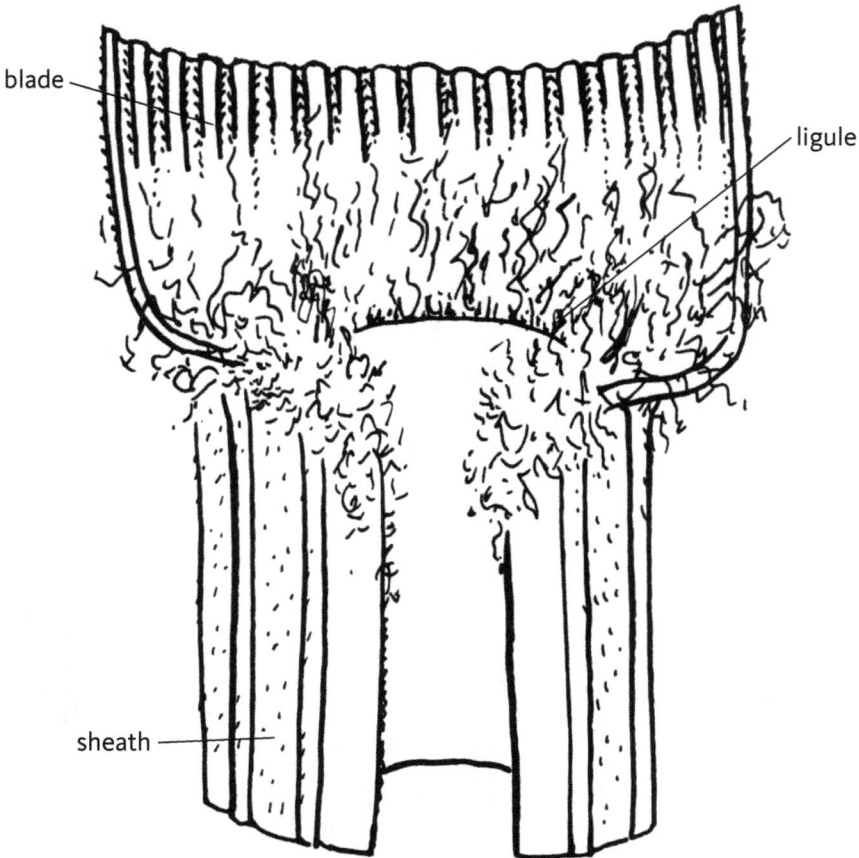

40

SYNONYM *Pleuraphis rigida* Thurb.

KEY CHARACTERS Large, erect bunchgrass. New growth sprouts at nodes. Ligule, nodes, collar, sheath and culms woolly.

VERNATION Curled. BLADES Flat at base, upper part rolled, erect, narrow, pointed; smooth, stiff; veins each side midrib 2-4; ribs prominent ventrally, 6-8; margin toothed; midrib prominent dorsally; width 1-2 mm; length 4-10 cm. AURICLE None. LIGULE Hairy to woolly, small, 1 mm. COLLAR Woolly. SHEATH Round, margin woolly. ROOTS Hard, scaly, rhizomatous. CULM Round, branched.

HABITAT Deserts, open juniper woodlands.

Hopia obtusa (Kunth) F. Zuloaga & O. Morrone
VINE-MESQUITE

Hopia obtusa (Kunth) F. Zuloaga & O. Morrone
VINE-MESQUITE

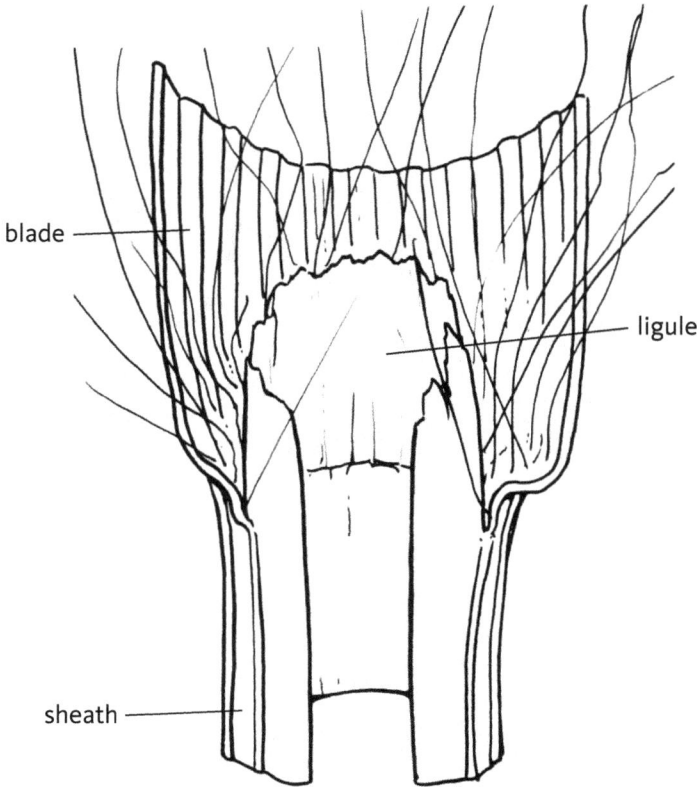

SYNONYM *Panicum obtusum* Kunth

KEY CHARACTERS Erect, almost a sod-former, strongly stoloniferous; stolons up to 10 feet long. Blades flat, wide, long, stiff, narrow, pointed. Ligule membranous also long hairs 3–5 mm. Sheath with glandular hairs at base of plant.

VERNATION Curled. **BLADES** Flat, wide, long; glandular hairs ventrally; veins each side of midrib 4; ribs prominent; margin toothed; midrib prominent dorsally; width 4–6 mm, length 10–20 cm. **AURICLE** None. **LIGULE** Membranous, with occasional hairs 1–3 mm, obtuse-notched. **COLLAR** Margin long-hairy. **SHEATH** Glandular hairs at base of plant. **NODE** Glabrous (stolons hairy). **INTERNODE** Glabrous. **ROOTS** Stoloniferous. **CULM** Elliptical, not branched.

HABITAT Seasonally wet sand or gravel, especially streambanks, ditches, and wet pastures.

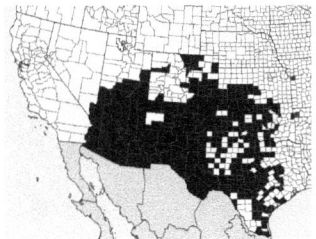

42 *Koeleria macrantha* (Ledeb.) J.A. Schultes
JUNEGRASS

Koeleria macrantha (Ledeb.) J.A. Schultes
JUNEGRASS

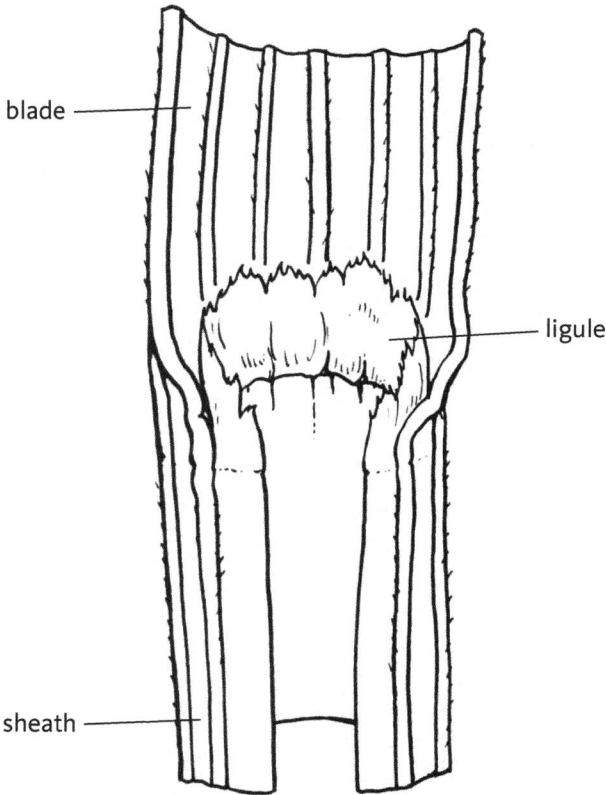

SYNONYM *Koeleria cristata* (L.) Pers.

KEY CHARACTERS Small, erect bunchgrass. Blade flat, conspicuously ribbed ventrally, soft. Internode canescent near node. Ligule membranous, truncate-ciliate, 1/2 mm.

VERNATION Folded. **BLADES** Flat, drooping, narrow, pointed; rough ventrally, canescent dorsally, soft; veins each side of midrib usually 2; ribs prominent ventrally; margin toothed; width 2 mm, length 5–15 cm. **AURICLE** None. **LIGULE** Membranous, small, 1/4–1/2 mm, truncate-ciliate. **COLLAR** Glabrous to canescent. **SHEATH** Usually canescent, round. **NODE** Usually glabrous. **INTERNODE** Usually canescent above node. **ROOTS** Fibrous. **CULM** Round, not branched.

HABITAT Semi-arid to mesic habitats such as dry prairies and grassy woods; soils usually sandy.

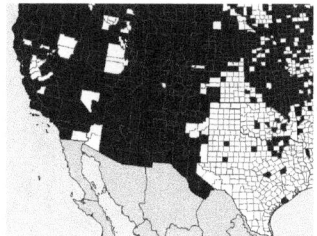

Leptochloa dubia (Kunth) Nees
GREEN SPRANGLETOP

Leptochloa dubia (Kunth) Nees
GREEN SPRANGLETOP

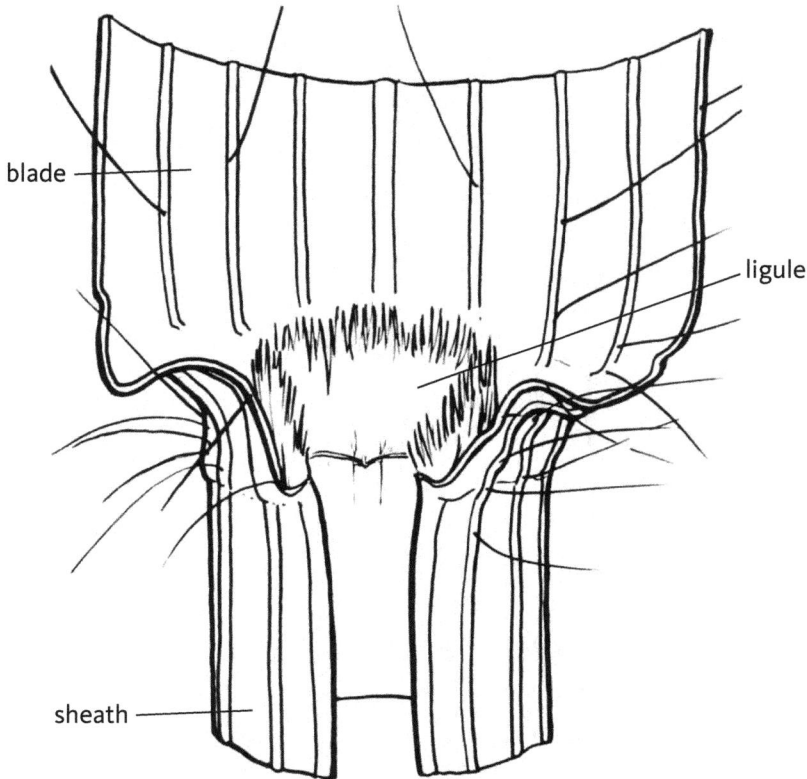

SYNONYM *Disakisperma dubium* (Kunth) P.M. Peterson & N. Snow

KEY CHARACTERS Erect, medium bunchgrass. Collar margin hairy. Culms flat. Ligule hairy, small, 1/2 mm, occasional hairs 3–5 mm long. Blades flat, wide, long drooping. Sheath veined, margin papery.

VERNATION Folded (reported in Colorado as curled). **BLADES** Flat, wide, long, drooping; occasional hairs dorsally 2–5 mm, on lower 1/3; veins on each side of midrib 2–3; ribs ventral and dorsal; margin toothed; midrib prominent ventrally; width 4–5 mm, length 15–25 cm. **AURICLE** None. **LIGULE** Hairy, small, 1/2 mm, margin hairs 3–5 mm. **COLLAR** Margin hairy, 3–5 mm. **SHEATH** Glabrous, veined, margin papery. **NODE** Glabrous. **INTERNODE** Glabrous. **ROOTS** Fibrous. **CULM** Flat, not branched.

HABITAT Well-drained, sandy or rocky soils.

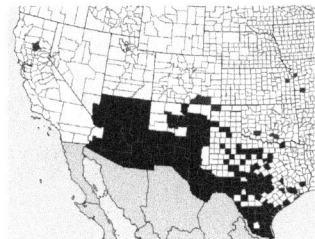

44 *Lolium perenne* L.
PERENNIAL RYEGRASS

Lolium perenne L.
PERENNIAL RYEGRASS

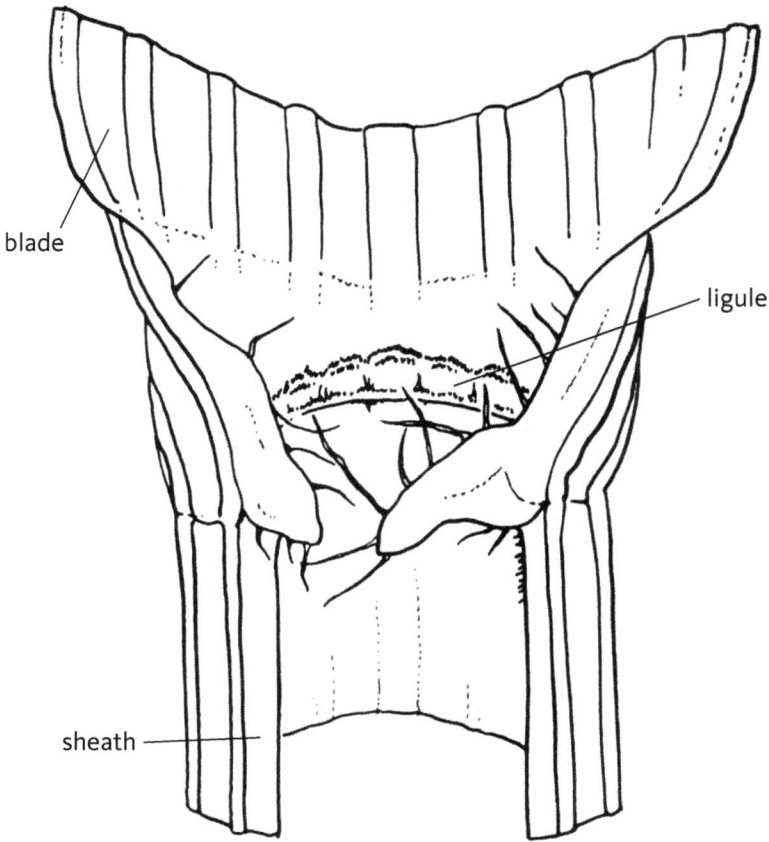

KEY CHARACTERS Introduced, small, erect to semi-erect bunchgrass. Blades glossy ventrally. Auricle present clawed or reduced. Sheath usually closed and pinkish at base.

VERNATION Folded. **BLADES** Flat to folded, wide, long drooping; glabrous; glossy dorsally; veins each side of midrib 2–3; ribs prominent ventrally; margin glabrous; midrib prominent dorsally; width 3–4 mm, length 8–15 cm. **AURICLE** Small, 1 mm.

LIGULE Membranous, 1–2 mm, acute-toothed. **COLLAR** Glabrous, divided. **SHEATH** Usually closed, strongly ribbed, glabrous. **NODE** Glabrous. **INTERNODE** Glabrous. **ROOTS** Fibrous. **CULM** Elliptical.

HABITAT Disturbed areas; often included in pasture seed mixes.

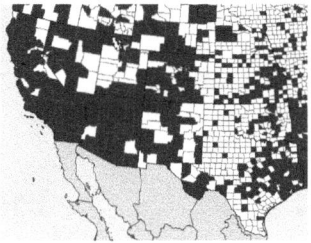

45 *Lycurus phleoides* Kunth
WOLFTAIL

Lycurus phleoides Kunth
WOLFTAIL

blade

ligule

sheath

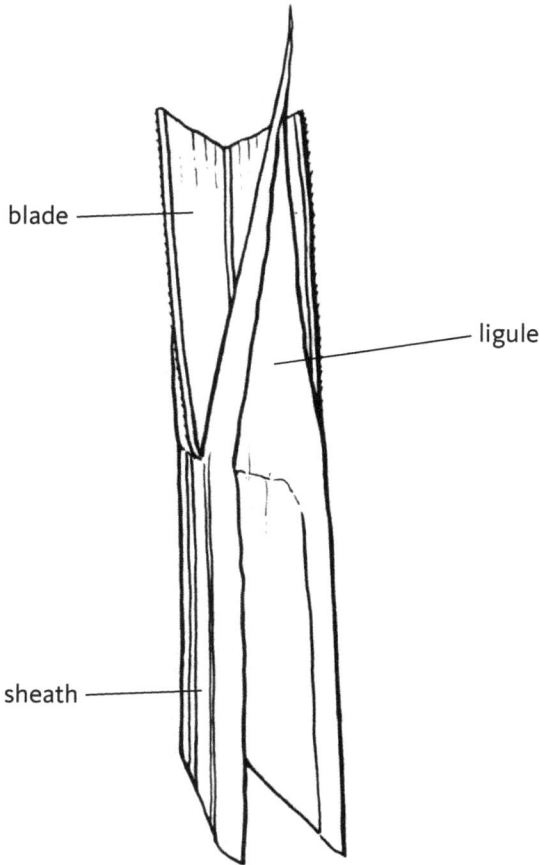

SYNONYM *Muhlenbergia phleoides* (Kunth) Columbus

KEY CHARACTERS Small, erect bunchgrass. Blades and sheath conspicuous white margin, flattened at base. Ligule large acute. Culms branch at nodes.

VERNATION Folded. **BLADES** Usually folded, drooping, glabrous; veins each side of midrib 2; margin white, toothed; ribs indistinct; midrib prominent dorsally; width 1 mm, length 3-10 cm. **AURICLE** None. **LIGULE** Membranous, large, 2-5 mm, acute-entire. **COLLAR** Glabrous. **SHEATH** Flat, margin hyaline white. **NODE** Short canescent. **INTERNODE** Glabrous to short canescent near nodes. **ROOTS** Fibrous. **CULM** Flat, branched.

HABITAT Rocky, open slopes.

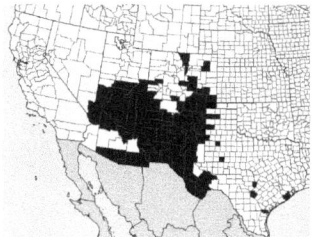

46 *Muhlenbergia curtifolia* Scribn.
UTAH MUHLY

UTAH MUHLY

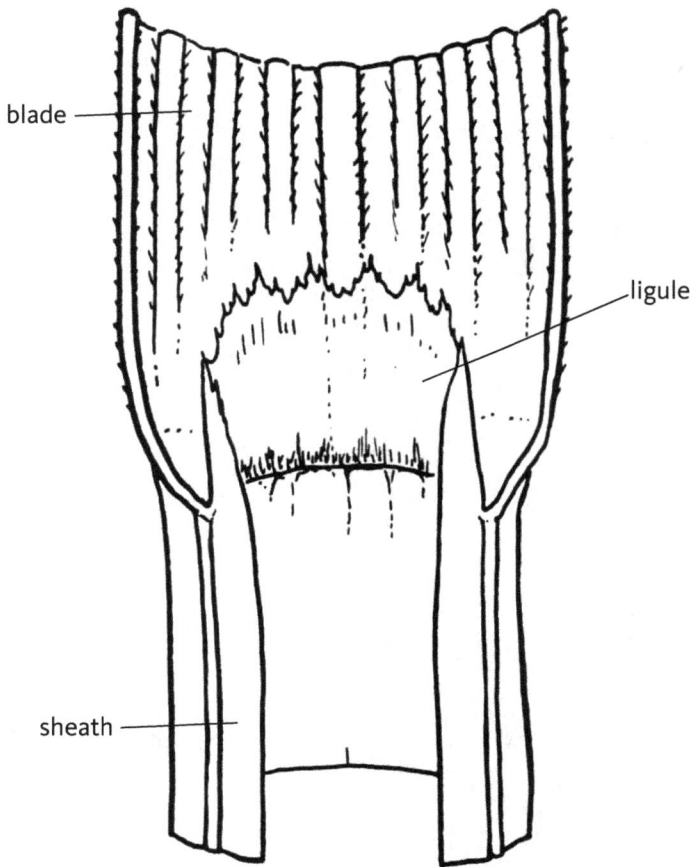

KEY CHARACTERS Erect, medium-sized bunchgrass. Ligule membranous, 1–2 mm, acute-lacerate. Blades rolled, long and narrow; blade veins 3–4. Plant glabrous.

VERNATION Clasped. **BLADE** Rolled, semi-erect, pointed; rough, stiff, pubescent ventrally; veins each side of midrib 3–4; ribs prominent dorsally; margin toothed; midrib prominent ventrally; width 1 mm, length 1–2½ cm. **AURICLE** None. **LIGULE** Membranous, small, 1–2 mm, acute-lacerate. **COLLAR** Glabrous. **SHEATH** Glabrous, veined, round. **NODE** Glabrous. **INTERNODE** Pubescent to glabrous. **ROOTS** Fibrous. **CULM** Round, not branched.

HABITAT Moist cracks and ledges of cliffs, and near large calcareous boulders in canyons.

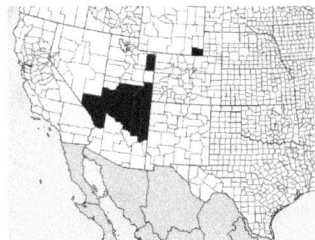

Muhlenbergia emersleyi Vasey
BULLGRASS

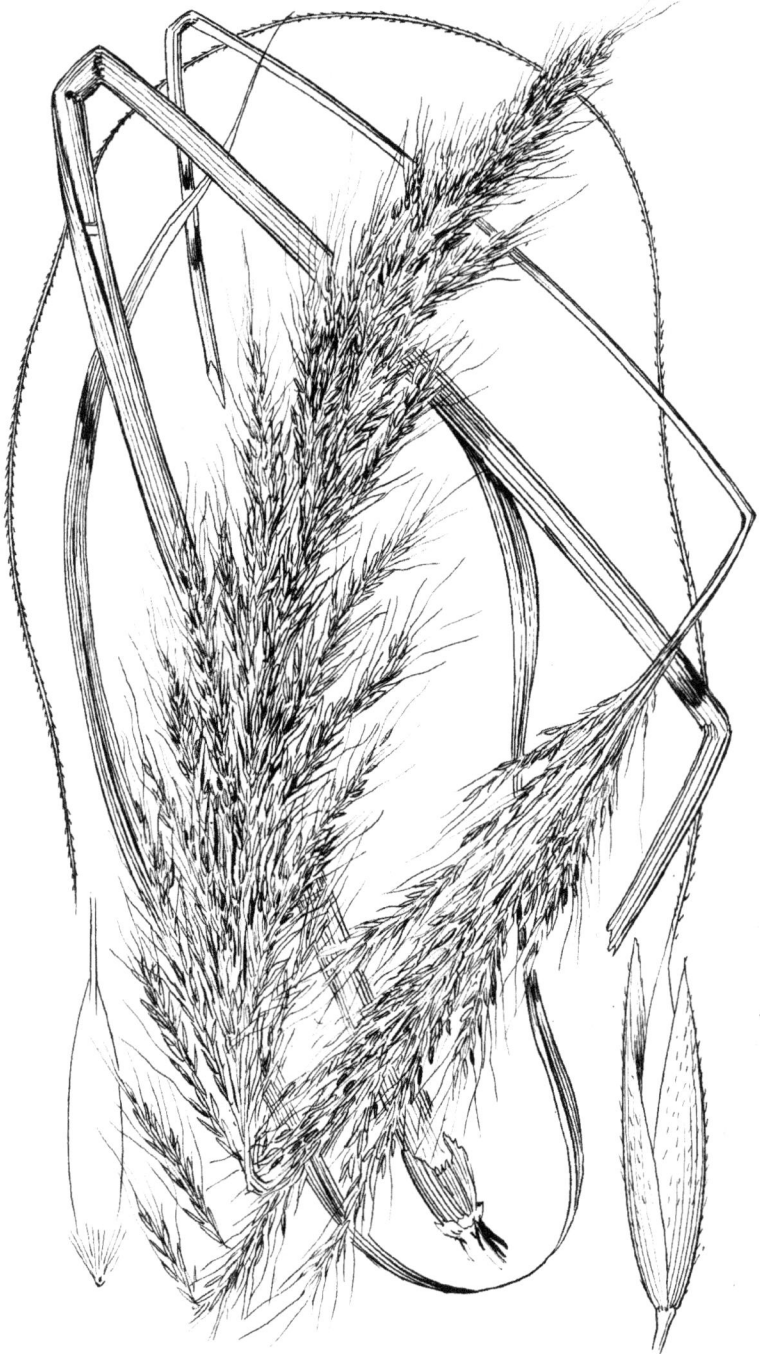

Muhlenbergia emersleyi Vasey
BULLGRASS

blade

ligule

sheath

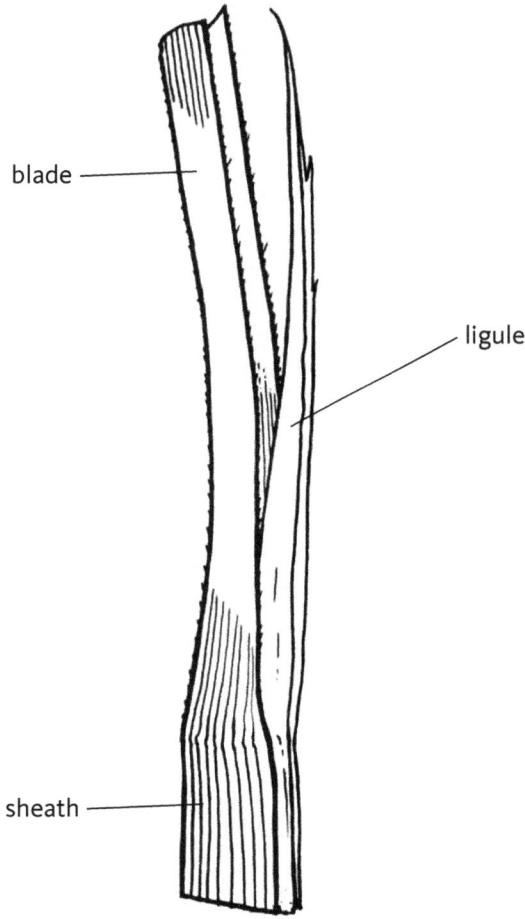

KEY CHARACTERS Erect, medium to large bunchgrass. Ligule hyaline, 8-15 mm acute-lacerate. Blade folded, long, narrow.

VERNATION Folded. **BLADES** Usually folded, narrow, long; rough dorsally, stiff; veins each side of midrib 3-4; ribs prominent, numerous; margin toothed, midrib prominent, dorsally; width 2-3 mm, length 15-35 cm. **AURICLE** None. **LIGULE** Membranous, 8-15 mm, acute-lacerate. **COLLAR** Glabrous. **SHEATH** Glabrous, veined. **NODE** Glabrous. **ROOTS** Fibrous. **CULM** Elliptical, not branched.

HABITAT Rocky slopes, gravelly washes, canyons, cliffs, and arroyos; soils often derived from limestone.

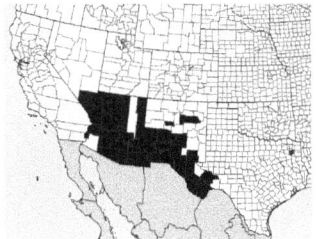

48 *Muhlenbergia montana* (Nutt.) A.S. Hitchc.
MOUNTAIN MUHLY

Muhlenbergia montana (Nutt.) A.S. Hitchc.
MOUNTAIN MUHLY

48

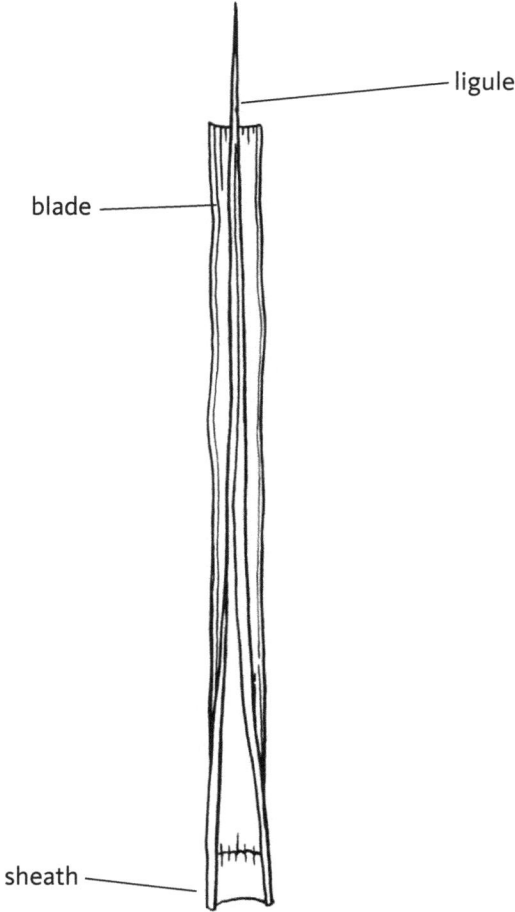

ligule

blade

sheath

KEY CHARACTERS Erect to semi-erect, small bunchgrass. Blade rolled, 5–20 mm, narrow. Ligule long, hyaline, often shredded, 5–8 mm.

VERNATION Folded (reported in Colorado as clasped). **BLADES** Rolled, narrow, pointed; glabrous and soft; margin toothed; width 1/4–1/2 mm, length 5–20 cm. Mature blades may be flat. **AURICLE** None. **LIGULE** Membranous, large, 6–12 mm, acute-entire. **COLLAR** Distinct, entire. **SHEATH** Glabrous, margin light color. **NODE** Glabrous. **INTERNODE** Glabrous. **ROOTS** Fibrous. **CULM** Round, not branched.

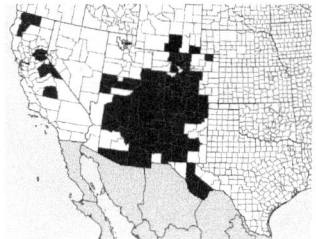

HABITAT Rocky slopes and ridges; dry meadows and open grasslands.

Muhlenbergia porteri Scribn. ex Beal

BUSH MUHLY

Muhlenbergia porteri Scribn. ex Beal
BUSH MUHLY

blade

ligule

sheath

....

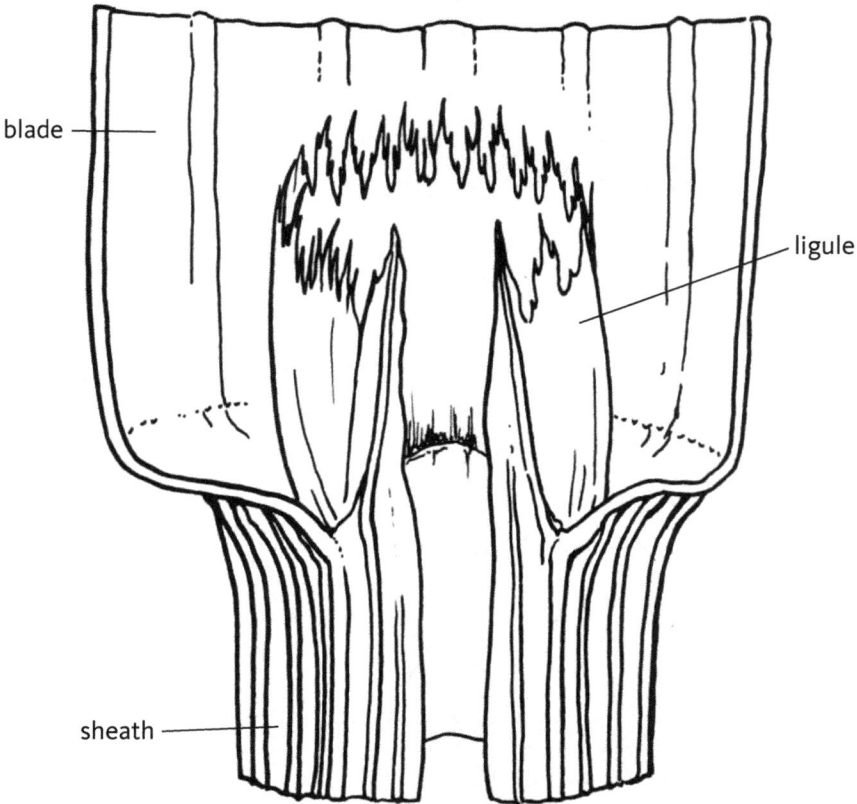

KEY CHARACTERS Large, decumbent bunchgrass. Ligule conspicuous, truncate-lacerate, 1–2 mm. Blade short, flat. Branches at nodes. Usually found clambering over shrubs. Culms small, long and semi-woody, persistently green. Florets purple-awned.

....

VERNATION Curled. **BLADES** Flat, narrow, short, pointed; rough dorsally; veins each side of midrib 2; ribs not prominent; margin toothed, midrib not prominent; width 1 mm, length 3–8 cm. **AURICLE** None. **LIGULE** Membranous, 1–2 mm, hyaline, truncate-lacerate. **COLLAR** Glabrous. **SHEATH** Open, glabrous, veined. **NODE** Glabrous. **INTERNODE** Glabrous. **ROOTS** Fibrous. **CULM** Elliptical to round, branched.

....

HABITAT Among boulders on rocky slopes and on cliffs, and in dry arroyos, desert flats, and grasslands, frequently under shrubs.

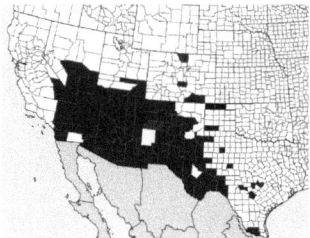

SANDHILL MUHLY

Muhlenbergia pungens Thurb. ex Gray
SANDHILL MUHLY

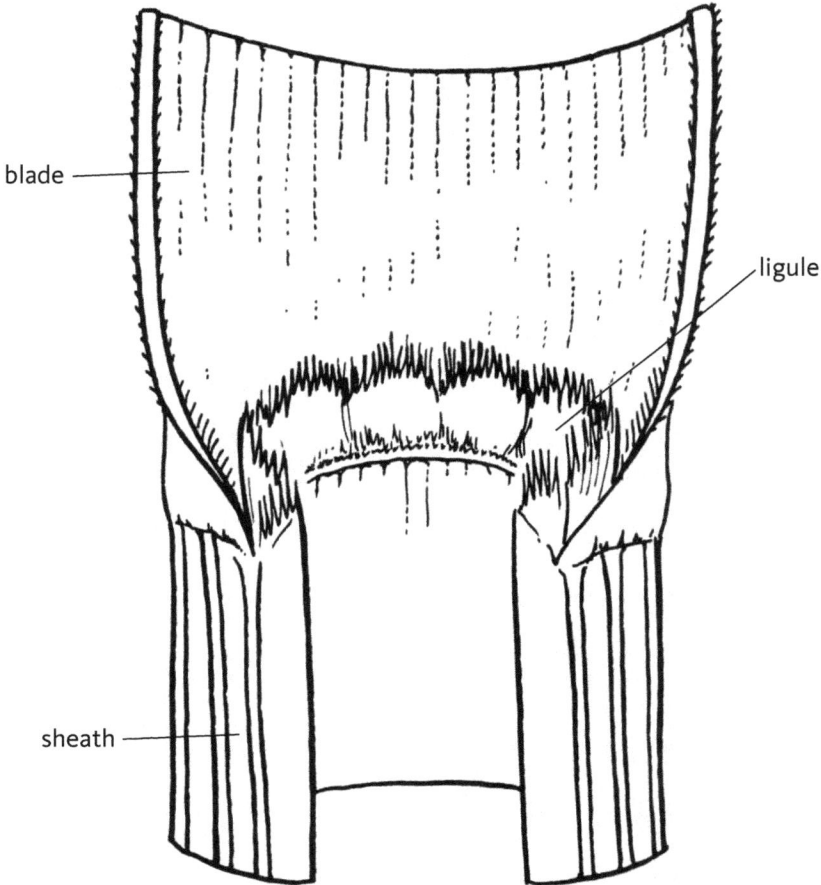

blade

ligule

sheath

KEY CHARACTERS Semi-erect, medium sod-former. Node and internode canescent. Blades rolled, pungent. Rhizomatous. Panicle open, reddish. Prefers sandy soil.

VERNATION Clasped. **BLADES** Rolled, very sharp pointed (pungent); rough ventrally, stiff; ribs indistinct; width 1 mm, length 3–5 cm. **AURICLE** None. **LIGULE** Membranous, small, 1/2 mm; truncate-ciliate. **COLLAR** Glabrous. **SHEATH** Open, glabrous, round, papery margined. **NODE** Canescent. **INTERNODE** Canescent. **ROOTS** Rhizomatous. **PANICLE** Reddish-brown. **CULM** Round.

HABITAT Loose sandy soils of slopes and near sand dunes; desert shrublands and open woodlands.

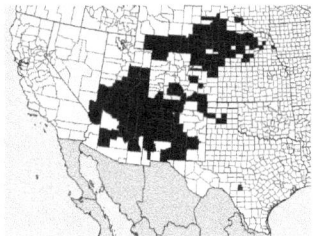

Muhlenbergia richardsonis (Trin.) Rydb.
MAT MUHLY

Muhlenbergia richardsonis (Trin.) Rydb.
MAT MUHLY

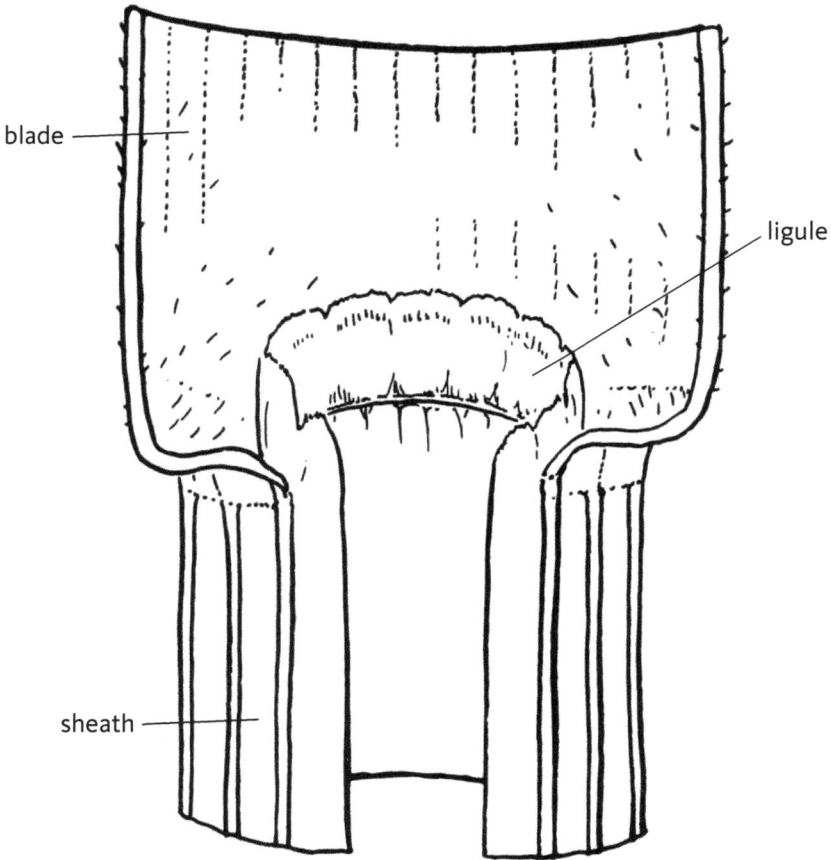

KEY CHARACTERS Erect to semi-erect, branched, forming a mat. Sheath margin white. Blades short, rolled, erect.

VERNATION Folded. **BLADES** Rolled, erect, narrow, pointed; glabrous, rough ventrally, stiff; margin toothed; width 1/2 mm, length 1–5 cm. **AURICLE** None. **LIGULE** Membranous, small, 1/2–1 mm, truncate-toothed. **COLLAR** Glabrous. **SHEATH** Glabrous, margin white hyaline, elliptical. **NODE** Glabrous. **INTERNODE** Glabrous. **ROOTS** Fibrous, short rhizomes. **CULM** Elliptical, branched.

HABITAT Open places in alkaline meadows, prairies, sandy arroyo bottoms, talus slopes, rocky flats, riverbanks.

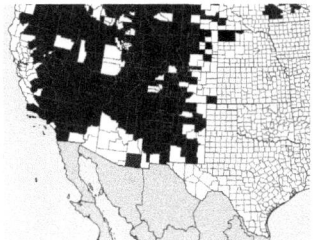

Muhlenbergia rigens (Benth.) A.S. Hitchc.
DEERGRASS

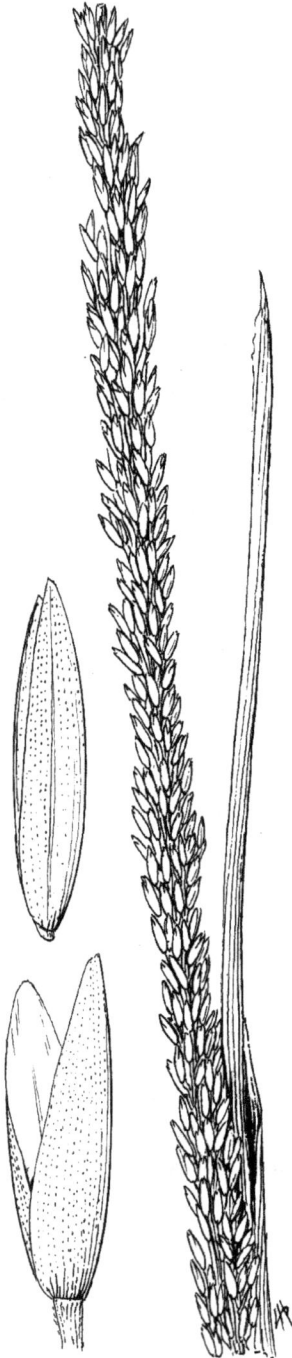

Muhlenbergia rigens (Benth.) A.S. Hitchc.

DEERGRASS

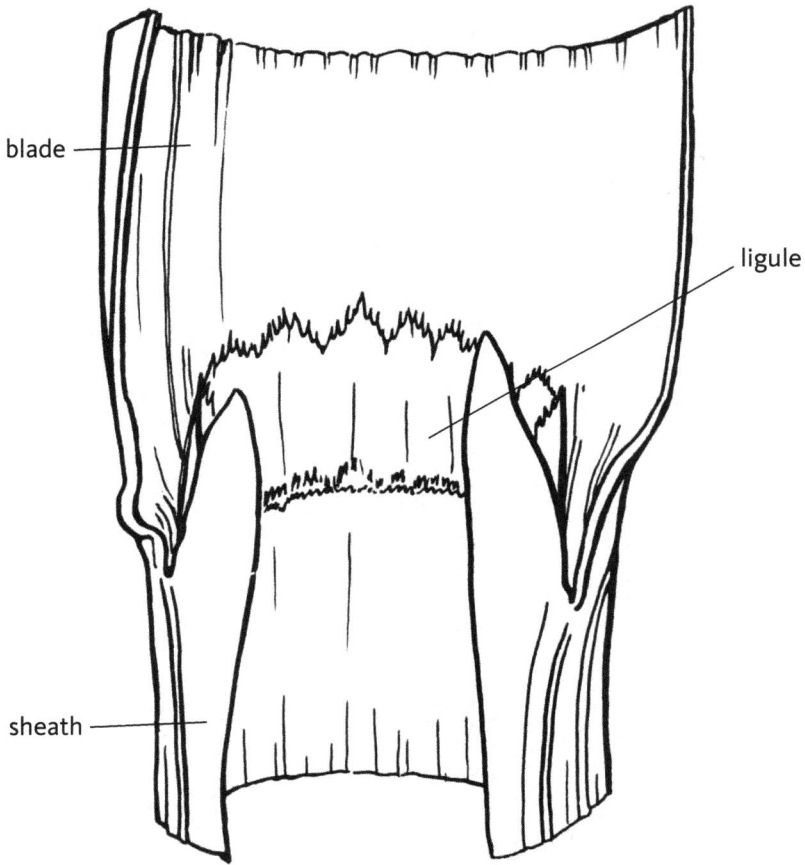

KEY CHARACTERS Large, erect, bunchgrass. Blades long, narrow, rolled to flat. Ligule membranous, large, acute-lacerate, 2–4 mm.

VERNATION Clasped. **BLADES** Flat to rolled, rough, pointed, veins each side of midrib 2; ribs prominent ventrally and dorsally; margin toothed; width 1–2 mm, length 10–20 cm. **AURICLE** None. **LIGULE** Membranous, 2–4 mm, obtuse-lacerate. **COLLAR** Glabrous. **SHEATH** White margin, round, glabrous. **NODE** Glabrous. **INTERNODE** Glabrous. **ROOTS** Fibrous. **CULM** Round, not branched.

HABITAT Sandy washes, gravelly canyon bottoms, rocky drainages, moist, sandy slopes; often along small streams; sometimes grown as an ornamental.

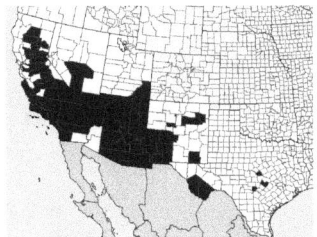

RING MUHLY

Muhlenbergia torreyi (Kunth) A.S. Hitchc. ex Bush
RING MUHLY

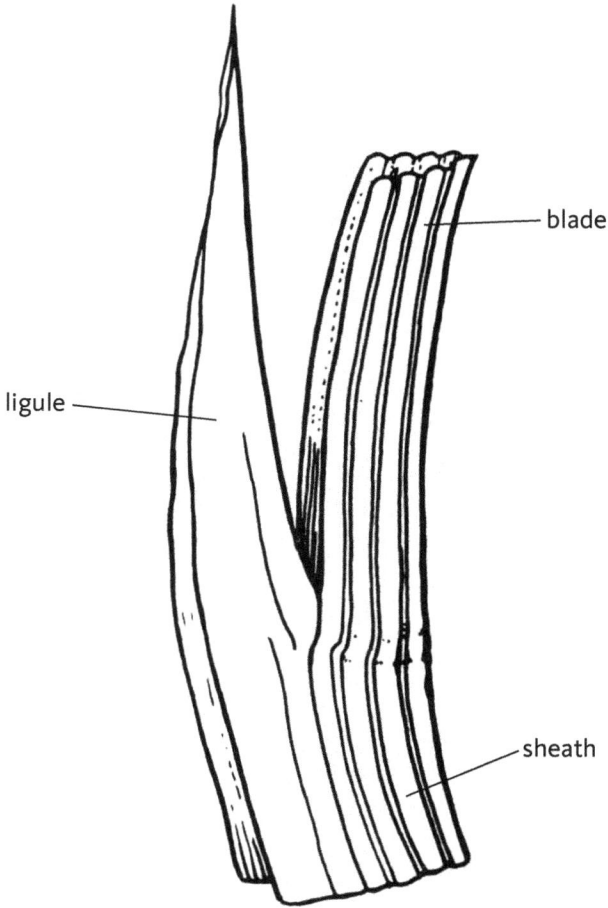

KEY CHARACTERS Semi-erect, forming dense mats, frequently open in center (ring-like). Blades and stems very small , flat. Ligule prominent.

VERNATION Folded. **BLADES** Small, folded, erect, narrow, pointed; glabrous; midrib not prominent; width 1/2 mm, length 1–2 cm. **AURICLE** None. **LIGULE** Membranous, medium, 2–3 mm, acute-entire. **COLLAR** Glabrous. **SHEATH** Glabrous, veined, round. **NODE** Canescent. **INTERNODE** Canescent. **ROOTS** Short rhizomes. **CULM** Flat, not branched, very small.

HABITAT Desert grasslands and open woodlands on sandy mesas, calcareous rock outcrops, and rocky slopes.

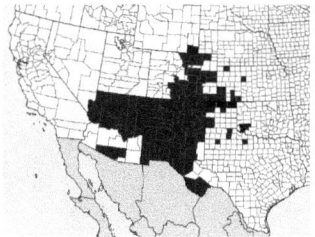

135

Pappophorum vaginatum Buckl.
FEATHER PAPPUSGRASS

Pappophorum vaginatum Buckl.
FEATHER PAPPUSGRASS

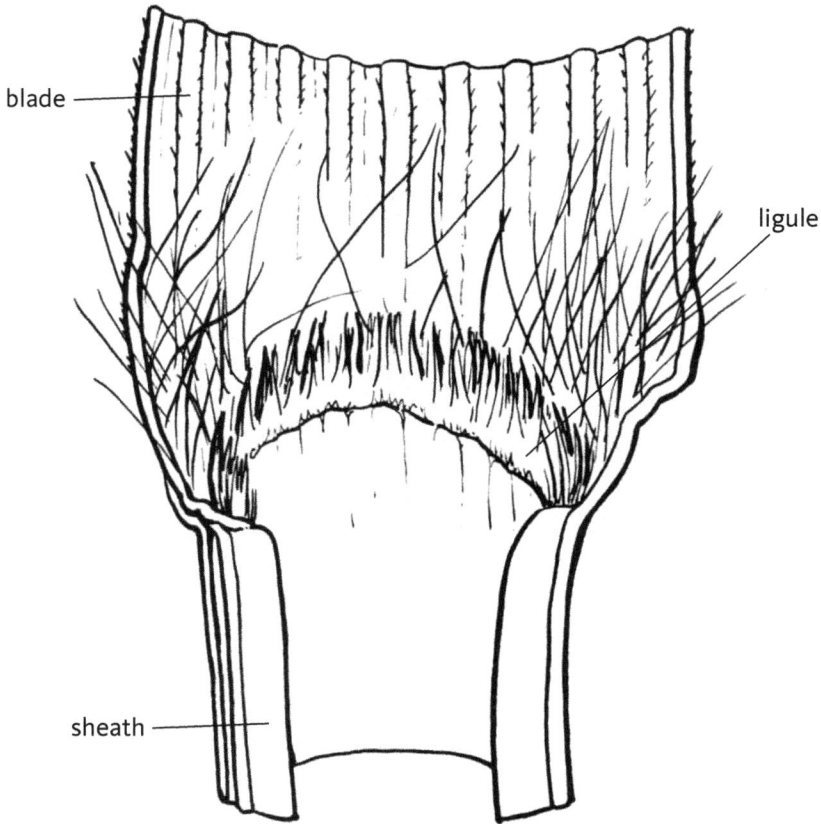

54

blade

ligule

sheath

SYNONYM *Pappophorum mucronulatum* Nees.

KEY CHARACTERS Large, erect bunchgrass. Culms branched at nodes. Ligule membranous 2-4 mm, truncate-ciliate. Blades flat, drooping, narrow, pointed, 15-30 cm long.

VERNATION Curled. **BLADES** Flat, drooping, narrow, pointed; rough ventrally, soft; veins each side of midrib 3-4; ribs prominent dorsally; margin toothed; midrib not prominent; width 3-4 mm, length 15-30 cm. **AURICLE** None. **LIGULE** Membranous, 2-4 mm, bearded, truncate-ciliate. **COLLAR** Glabrous. **SHEATH** Glabrous, round, margin paper-like. **NODE** Glabrous. **INTERNODE** Glabrous. **ROOTS** Fibrous. **CULM** Round, branched at nodes.

HABITAT Grassy plains, roadsides.

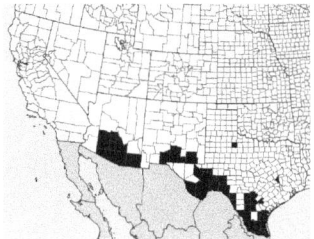

Pascopyrum smithii (Rydb.) A. Löve
WESTERN WHEATGRASS

Pascopyrum smithii (Rydb.) A. Löve

WESTERN WHEATGRASS

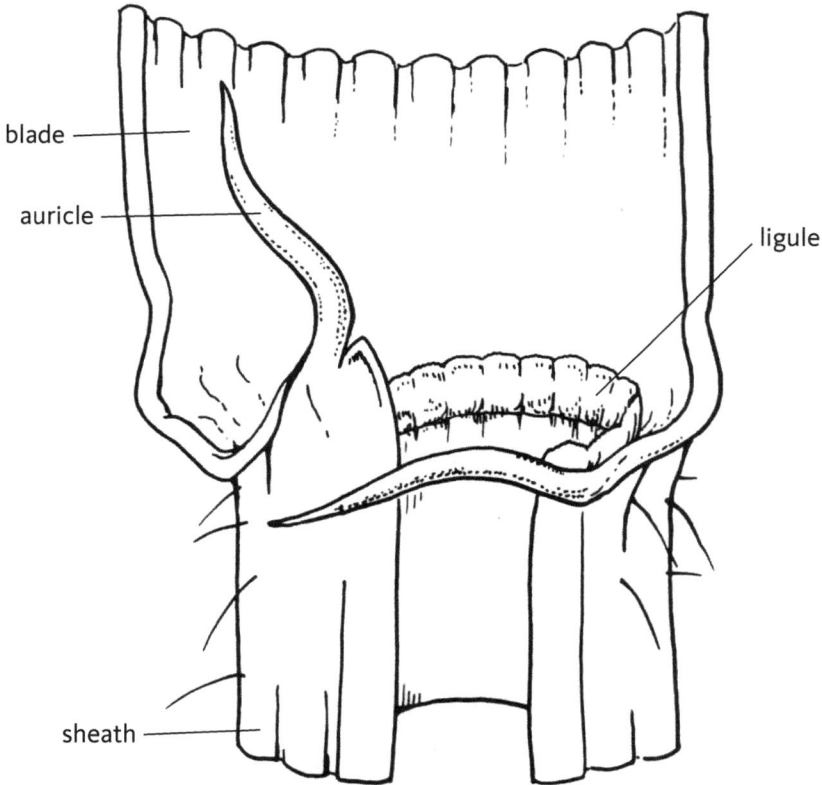

55

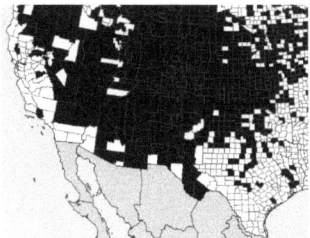

SYNONYM *Agropyron smithii* Rydb.

KEY CHARACTERS Erect, individual plants or forming a loose sod. Auricles small, blade ribs prominent dorsally, growth usually sod, blade color usually a blue-green.

VERNATION Curled. **BLADES** Usually rolled blue-green, erect, narrow pointed; rough dorsally, pubescent ventrally; veins each side of midrib, 6-8; ribs prominent ventrally; margin toothed, midrib not prominent; 2-3 mm wide, 5-15 cm long. **AURICLE** Small, 1/4-1/2 mm, usually brownish-red. **LIGULE** Collar-like (membranous), small, 1/4-1/2 mm; truncate-toothed. **COLLAR** Smooth, divided. **SHEATH** Smooth, round, veined; lower pinkish to red. **NODE** Glabrous. **INTERNODE** Glabrous. **ROOTS** Fibrous, rhizomatous. **CULM** Round, not branched.

HABITAT Sagebrush communities, mesic alkaline meadows.

56 *Piptochaetium pringlei* (Beal) Parodi
PRINGLE NEEDLEGRASS

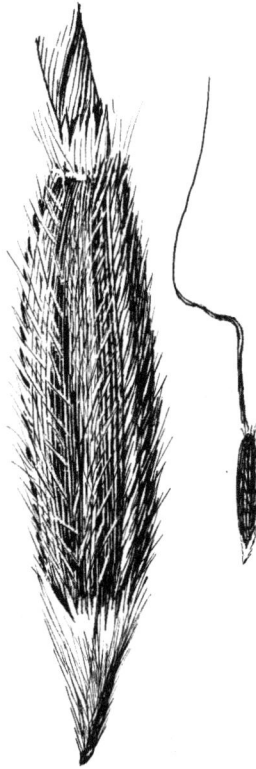

Piptochaetium pringlei (Beal) Parodi

PRINGLE NEEDLEGRASS

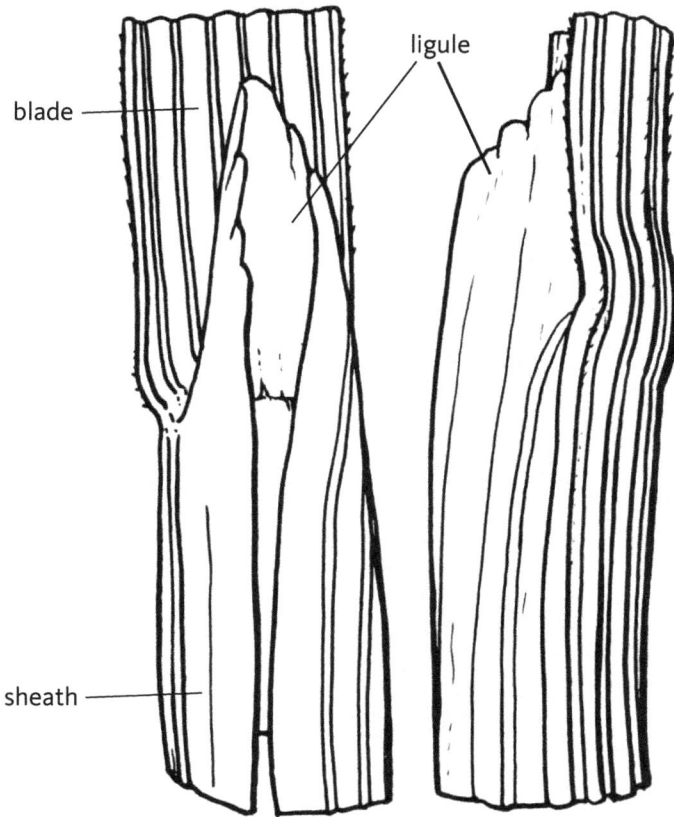

blade

ligule

sheath

- -

SYNONYM *Stipa pringlei* Scribn.

KEY CHARACTERS Small, erect bunchgrass. Blades flat, narrow, long. Ligule membranous, small 1–2 mm, obtuse-lacerate. Awns twisted, 1–3 cm.

VERNATION Curled. **BLADES** Flat, drooping, narrow, pointed; rough ventrally; veins each side of midrib 2; ribs prominent ventrally and dorsally; margin toothed; midrib prominent ventrally; width 2 mm, length 20–30 cm. **AURICLE** None. **LIGULE** Membranous, small, 1–2 mm, obtuse-lacerate. **COLLAR** Glabrous. **SHEATH** Elliptical, veined, margin hyaline. **NODE** Usually pubescent. **INTERNODE** Glabrous to short canescent near nodes. **ROOTS** Fibrous. **CULM** Elliptical, not branched.

HABITAT Oak woodlands, often where rocky.

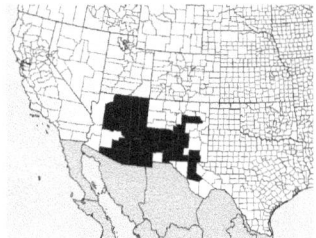

57 *Poa bigelovii* Vasey & Scribn.

BIGELOW BLUEGRASS

Poa bigelovii Vasey & Scribn.
BIGELOW BLUEGRASS

57

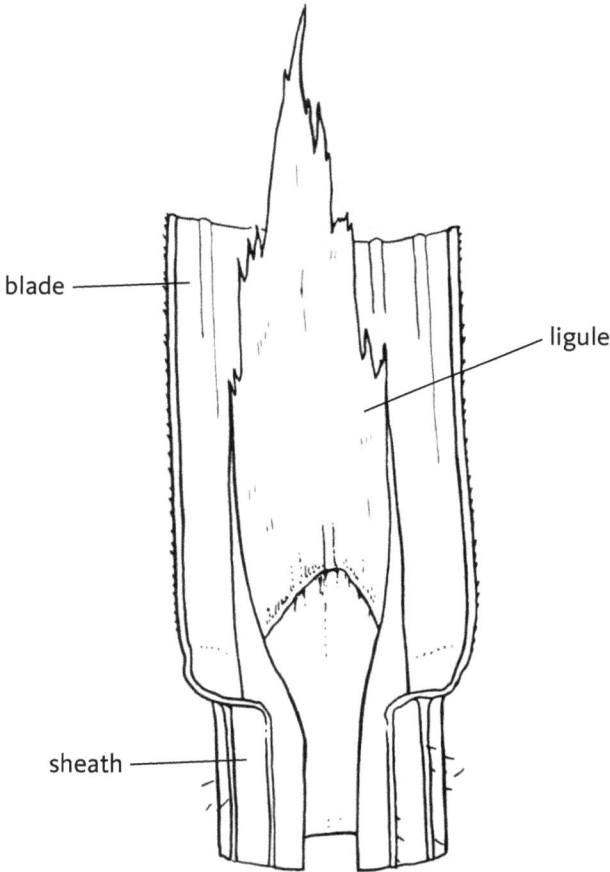

blade

ligule

sheath

KEY CHARACTERS Small annual, erect, bunchgrass. Plant glabrous. Ligule hyaline 2–4 mm, acute-lacerate. Blade flat, wide, drooping, long, narrow, pointed.

VERNATION Folded. **BLADES** Flat, wide, drooping, narrow, pointed, toothed; glabrous; veins each side of midrib 2; midrib not prominent; margin toothed; ribs not prominent; width 2–4 mm, length 4–12 cm, the tip boat-shaped. **AURICLE** None.
LIGULE Membranous, hyaline, 2–4 mm, acute-lacerate. **COLLAR** Glabrous, divided. **SHEATH** Glabrous, margin hyaline. **NODE** Glabrous. **INTERNODE** Glabrous. **ROOTS** Fibrous, annual. **CULM** Elliptical, not branched.

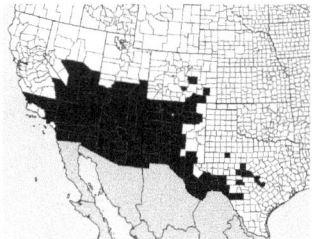

HABITAT Arid uplands, particularly on shaded and rocky slopes.

58 *Poa fendleriana* (Steud.) Vasey
MUTTON BLUEGRASS

Poa fendleriana (Steud.) Vasey

MUTTON BLUEGRASS

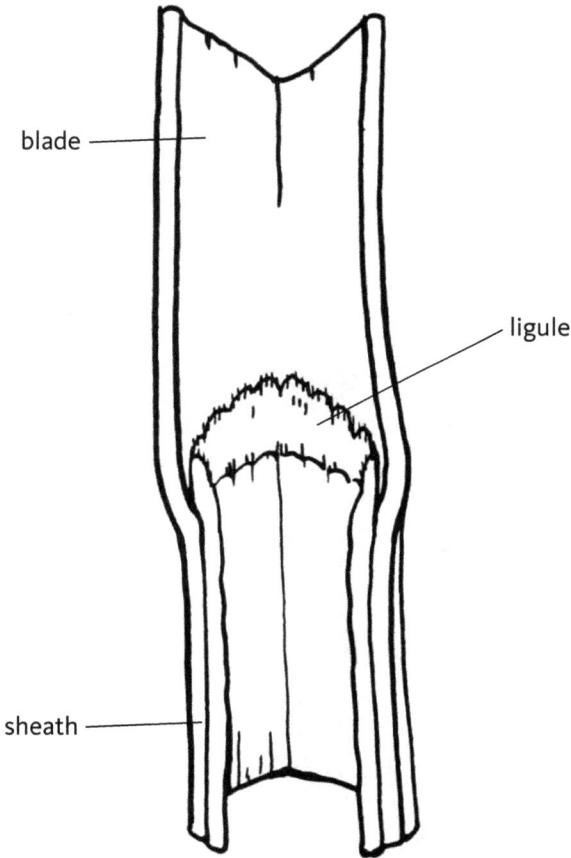

KEY CHARACTERS Small, erect bunchgrass. Blade long, narrow and folded. Ligule small, obtuse-toothed.

VERNATION Folded. **BLADES** Folded, erect, glabrous, stiff; ribs dorsal; margin glabrous; width 1/2–1 mm, length 10–20 cm, the tip boat-shaped. **AURICLE** None. **LIGULE** Membranous, small, 1/2 mm or less, obtuse-toothed. **COLLAR** Glabrous. **SHEATH** Short canescent to glabrous, elliptical, the margin hyaline. **NODE** Glabrous. **INTERNODE** Glabrous. **ROOTS** Fibrous. **CULM** Elliptical.

HABITAT Rocky slopes in sagebrush-scrub, chaparral, and high plains grasslands and forests; from low-elevations to lower alpine zone.

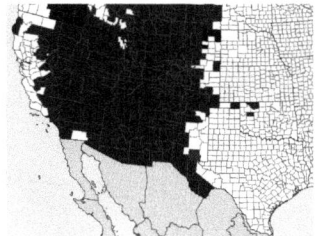

59 *Poa pratensis* L.
KENTUCKY BLUEGRASS

Poa pratensis L.
KENTUCKY BLUEGRASS

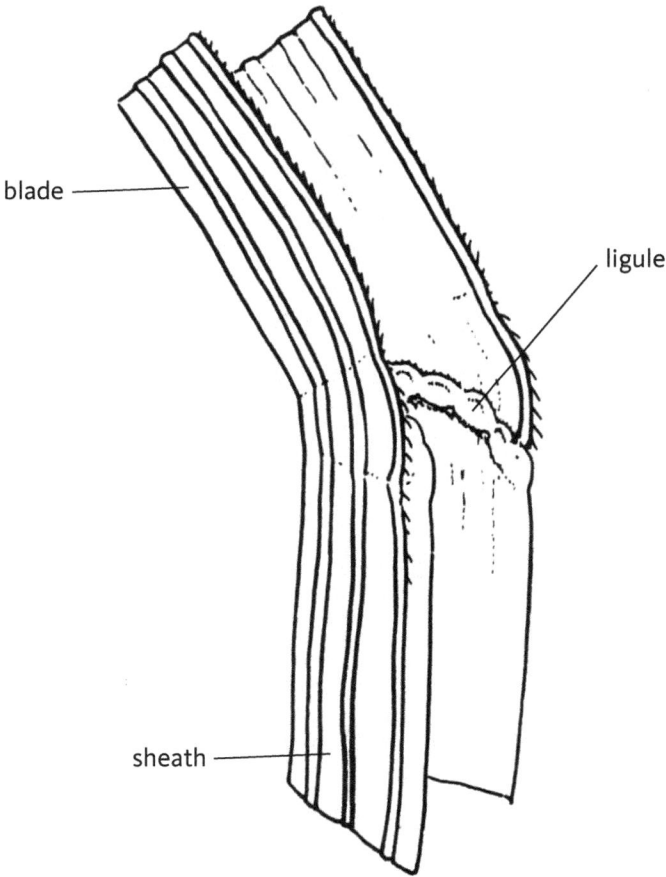

KEY CHARACTERS Naturalized; erect, sod-forminggrass. Blades long, ribbed, folded. Ligule very small, truncate-entire. Rhizomes short.

VERNATION Folded. **BLADES** Folded, soft; ribs prominent ventrally; margin smooth; width 1 mm, length 10-20 cm, the tip boat-shaped. **AURICLE** None. **LIGULE** Membranous, collarlike, small, 1/4-1/2 mm, truncate-entire. **COLLAR** Glabrous, not prominent. **SHEATH** Glabrous, veined, elliptical, with membranous edges. **NODE** Glabrous. **INTERNODE** Glabrous. **ROOTS** Short rhizomes. **CULM** Flat, not branched.

HABITAT In a wide range of moist to fairly dry habitats; commonly seeded in lawns.

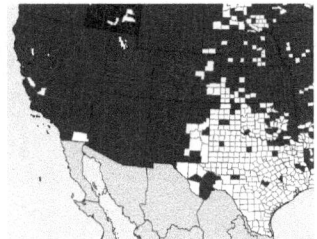

147

60 *Schedonnardus paniculatus* (Nutt.) Trel.
TUMBLEGRASS

Schedonnardus paniculatus (Nutt.) Trel.
TUMBLEGRASS

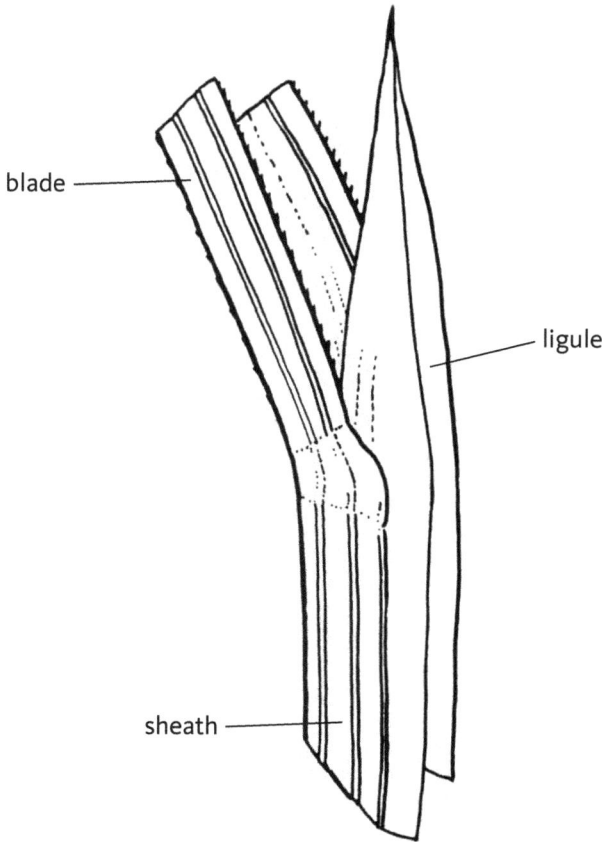

KEY CHARACTERS Small, erect bunchgrass. Blade and sheath with conspicuous white margin. Midrib prominent ventrally. Blade short, twisted. Ligule prominent, extending down sheath margin. At maturity, panicle breaks at base and acts as a tumbleweed, dispersing the seed.

VERNATION Folded. **BLADES** Folded, twisted, erect, narrow, pointed; smooth, soft; veins each side of midvein 1–2; ribs indistinct; margin white, toothed; midrib prominent dorsally; width 1 mm, length 2–3 cm. **COLLAR** Smooth. **AURICLE** None. **LIGULE** Membranous, 2–3 mm, acute-entire. **SHEATH** Flat, smooth, veined, margin paper-like, white. **NODE** Glabrous. **INTERNODE** Glabrous. **ROOTS** Fibrous. **CULM** Flat, not branched.

HABITAT Disturbed places.

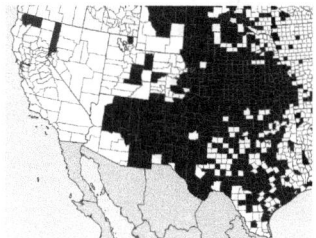

Schismus barbatus (Loefl. ex L.) Thellung

MEDITERRANEANGRASS

Schismus barbatus (Loefl. ex L.) Thellung

MEDITERRANEANGRASS

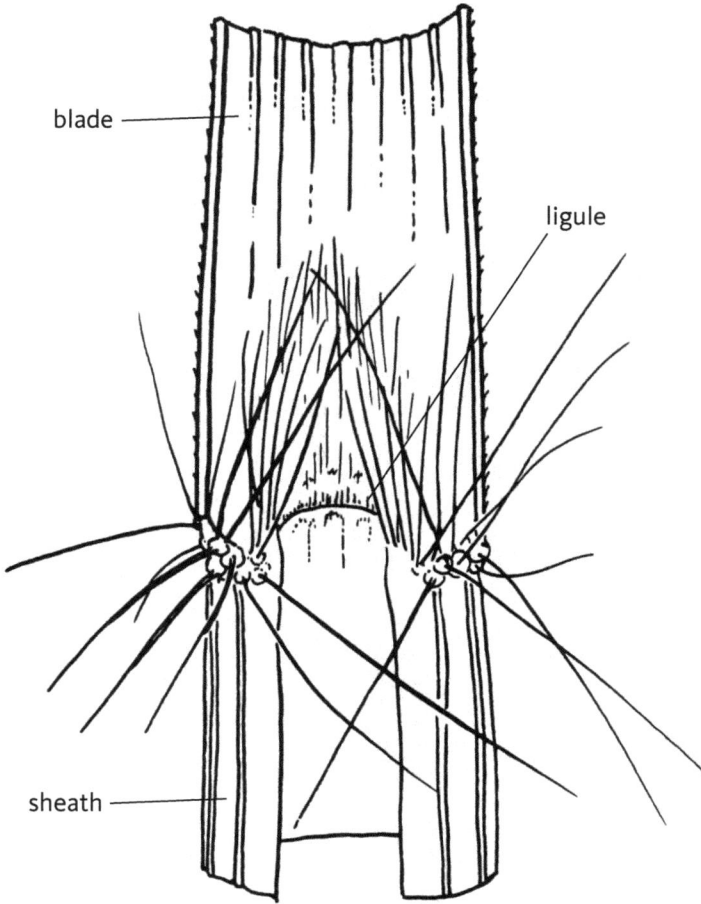

KEY CHARACTERS Introduced; semi-decumbent, small annual bunchgrass. Long hairs on ligule and collar. Blades rolled, very narrow and short. Roots annual.

VERNATION Curled. **BLADES** Rolled, narrow, short; hairy ventrally 3–8 mm; too small to determine veins and ribs; width 1/2–1 mm, length 3–8 cm. **AURICLE** None. **LIGULE** Hairy, small, 1–2 mm with scattered hairs 3–5 mm. **COLLAR** Hairs 1–5 mm. **SHEATH** Glabrous, veined, margin hyaline. **NODE** Glabrous. **INTERNODE** Glabrous. **ROOTS** Fibrous, annual. **CULM** Round, not branched.

HABITAT Sandy roadsides, disturbed fields, dry riverbeds.

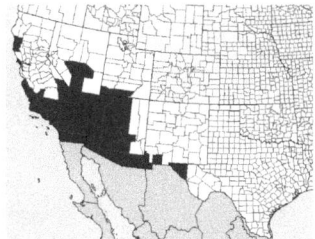

Schizachyrium cirratum (Hack.) Woot. & Standl.
TEXAS BLUESTEM

Schizachyrium cirratum (Hack.) Woot. & Standl.
TEXAS BLUESTEM

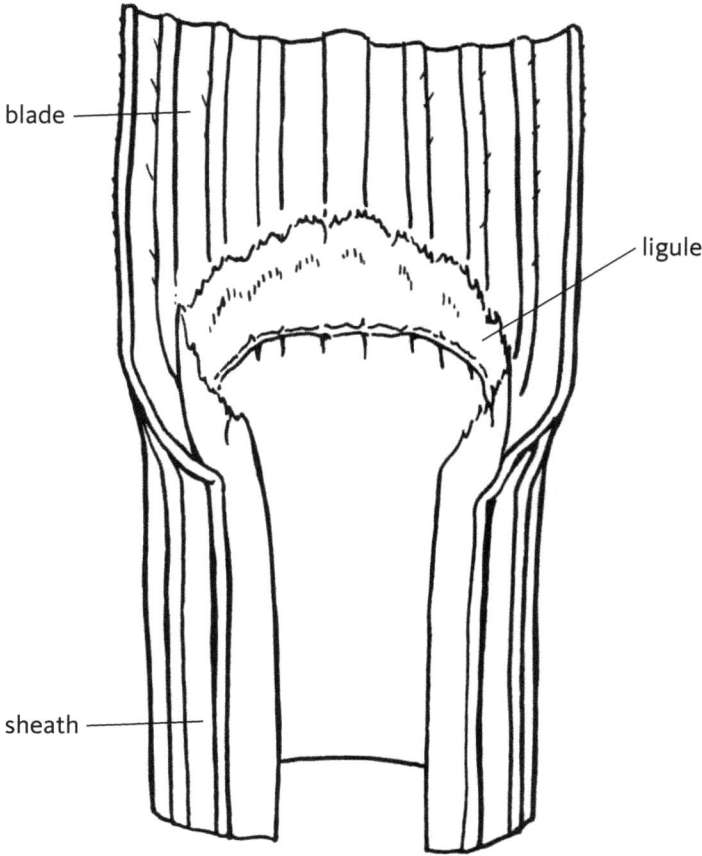

....

SYNONYM *Andropogon cirratus* Hack.

....

KEY CHARACTERS Small, erect bunchgrass. Old-growth blades usually flat; new-growth blades folded, long, narrow.

....

VERNATION Folded. **BLADES** Old, flat; new, folded; erect, narrow, pointed; veins each side of midrib 2-3; midrib prominent dorsally; 2-4 mm wide, 12-20 cm long. **AURICLE** None. **LIGULE** Membranous, small, 1-2 mm, obtuse-entire. **COLLAR** Glabrous. **SHEATH** Glabrous, round. **NODE** Glabrous. **INTERNODE** Glabrous. **ROOTS** Fibrous. **CULM** Round, sparingly branched.

....

HABITAT Rocky slopes at mid- to high-elevations.

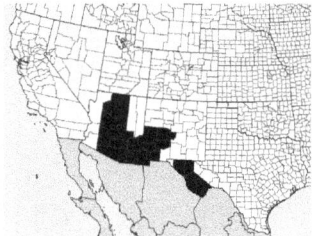

LITTLE BLUESTEM

Schizachyrium scoparium (Michx.) Nash
LITTLE BLUESTEM

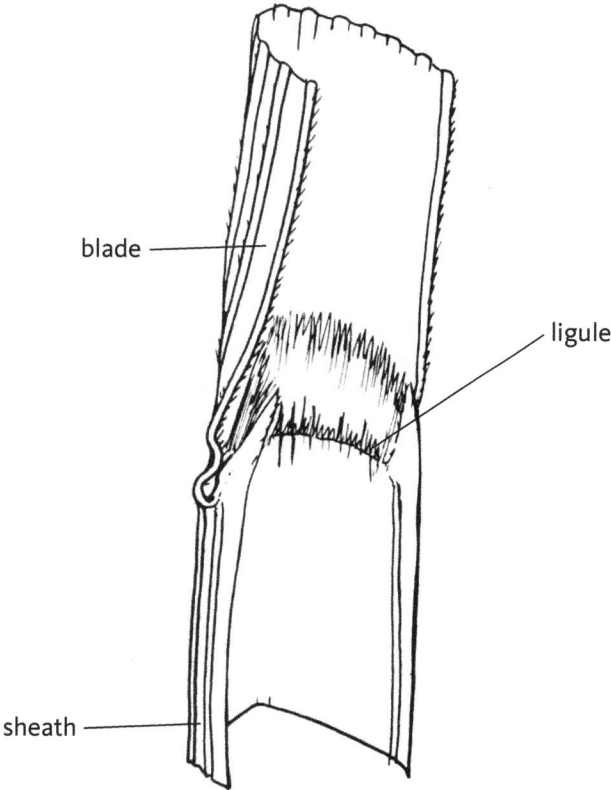

blade

ligule

sheath

....

SYNONYM *Andropogon scoparius* Michx.
....

KEY CHARACTERS Erect, medium to large bunchgrass. Blades usually folded, long and narrow, twisted. Stems elliptical to flat. Ligule very small and difficult to observe.
....

VERNATION Folded. **BLADES** Flat to folded, narrow, drooping, pointed; rough, margin toothed; veins each side of midrib usually 4; ribs prominent ventrally and dorsally; midrib prominent dorsally; 2 mm wide, 5–20 cm long. **AURICLE** None.

LIGULE Collar-like (membranous), small, 1/2 mm (occasionally to 2 mm), obtuse, finely lacerate. **COLLAR** Glabrous, very small. **SHEATH** Glabrous, flat, occasionally pinkish. **NODE** Glabrous, reddish-brown at maturity. **INTERNODE** Glabrous. **ROOTS** Fibrous. **CULM** Flat.
....

HABITAT Prairies, sandy meadows.

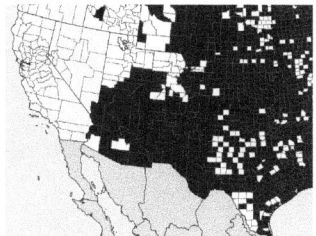

Setaria macrostachya Kunth
PLAINS BRISTLEGRASS

Setaria macrostachya Kunth
PLAINS BRISTLEGRASS

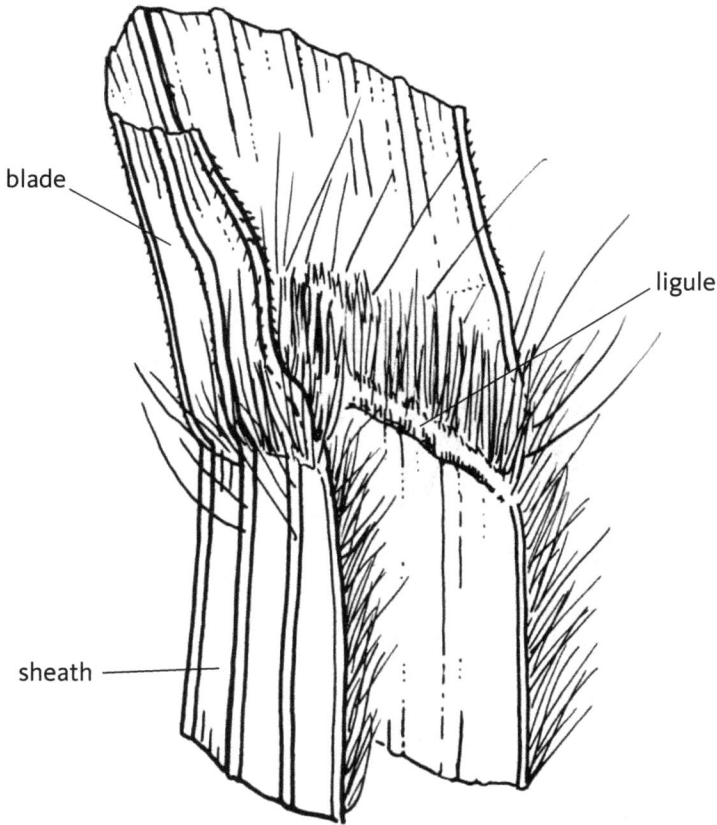

blade

ligule

sheath

KEY CHARACTERS Small, erect (lower semi-erect) bunchgrass. Culms flat. Blade in bud curled. Blade flat, long drooping. Ligule hairy, small, 1/2–1 mm. Sheath occasional marginal hairs.

VERNATION Curled. **BLADES** Flat, drooping, narrow, long; rough, soft, thin, veins each side of midrib 3–4; ribs prominent with lens; margin light color; midrib prominent ventrally; width 2–3 mm, length 5–20 cm. **AURICLE** None. **LIGULE** Hairy, small, 1/2–1 mm. **COLLAR** Margin hairy, divided. **SHEATH** Margin occasionally hairy. **NODE** Glabrous to hairy. **INTERNODE** Glabrous to hairy near nodes. **ROOTS** Fibrous. **CULM** Flat, not branched.

HABITAT Common in desert grasslands of the region, especially in southern parts of Arizona, New Mexico and Texas.

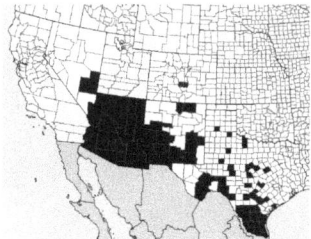

65 *Setaria parviflora* (Poir.) Kerguélen
KNOTROOT BRISTLEGRASS

Setaria parviflora (Poir.) Kerguélen

KNOTROOT BRISTLEGRASS

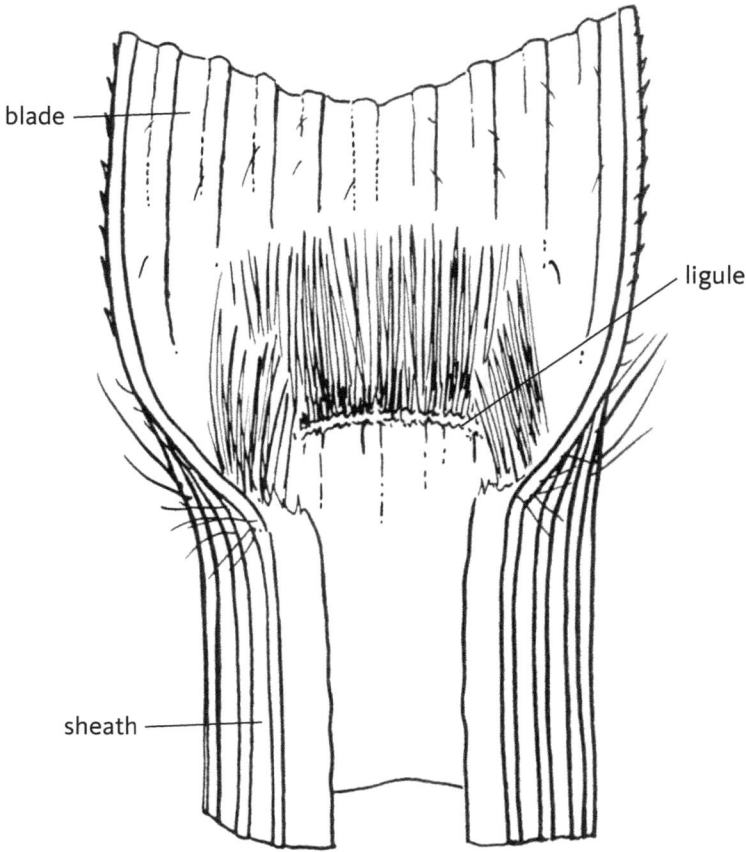

SYNONYM *Setaria geniculata* (Lam.) Beauv.

KEY CHARACTERS Large, erect bunchgrass. Ligule, node and lower internode hairy. Sheath margin hairy. Collar hairy. Blades folded.

VERNATION Curled. **BLADES** Folded, wide, long, drooping; hairy ventrally near ligule, these occasionally glandular; veins each side of midrib 4; ribs prominent; margin toothed; midrib not prominent; width 3-5 mm, length 10-20 cm. **AURICLE** None. **LIGULE** Hairy, 1-2 mm, bearded. **COLLAR** Hairy. **SHEATH** Margin and lower nodes, culm hairy. **NODE** Hairy. **INTERNODE** Hairy. **ROOTS** Fibrous. **CULM** Elliptical, branched.

HABITAT Moist places.

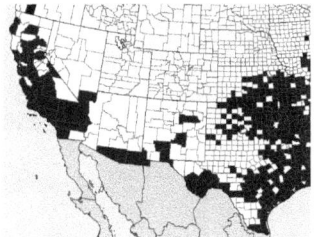

66 *Sorghum halepense* (L.) Pers.
JOHNSONGRASS

Sorghum halepense (L.) Pers.
JOHNSONGRASS

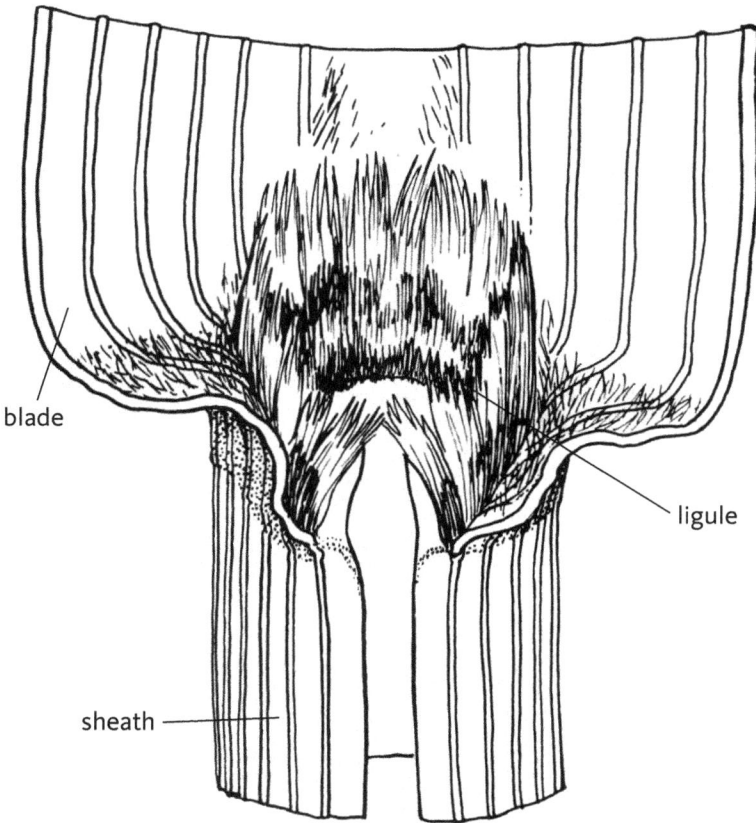

blade

ligule

sheath

....

KEY CHARACTERS Introduced and weedy; erect, sod-former. Massive rootstalks, rhizomes. Blades large, flat; blade margin white. Prefers wet areas, roadsides, irrigation ditches.

....

VERNATION Curled. **BLADES** Flat, wide, drooping, narrow, pointed; veins each side of midrib 5–7, ribs indistinct; margin toothed, white; midrib prominent dorsally; width 10–15 mm, length 20–40 cm. **AURICLE** None. **LIGULE** Hairy, obtuse, 1–2 mm. **COLLAR** Pubescent to glabrous. **SHEATH** Elliptical, veined. **NODE** Glabrous to very short canescent. **INTERNODE** Glabrous. **ROOTS** Rootstalk rhizomatous. **CULM** Elliptical, occasionally branched.

....

HABITAT Weedy in moist pastures and fields, roadsides, ditches, streambanks.

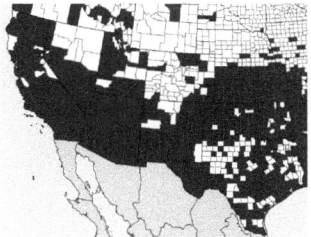

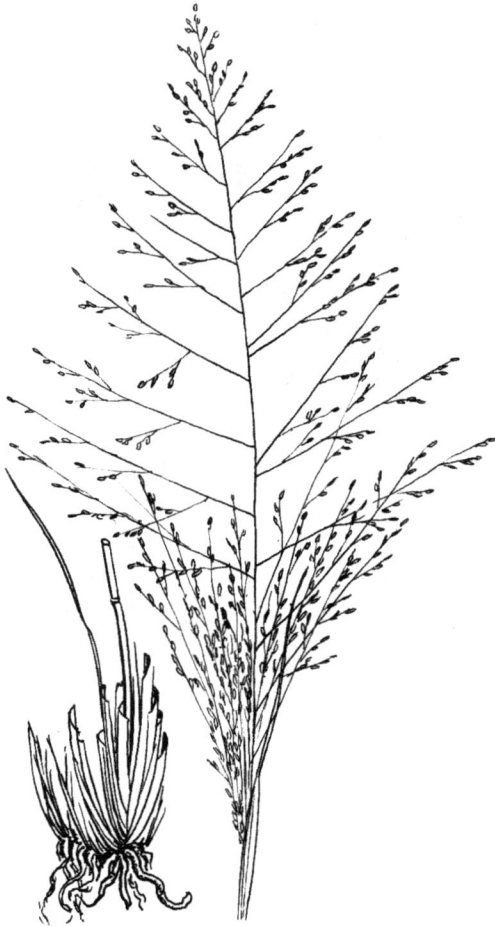

Sporobolus airoides (Torr.) Torr.
ALKALI SACATON

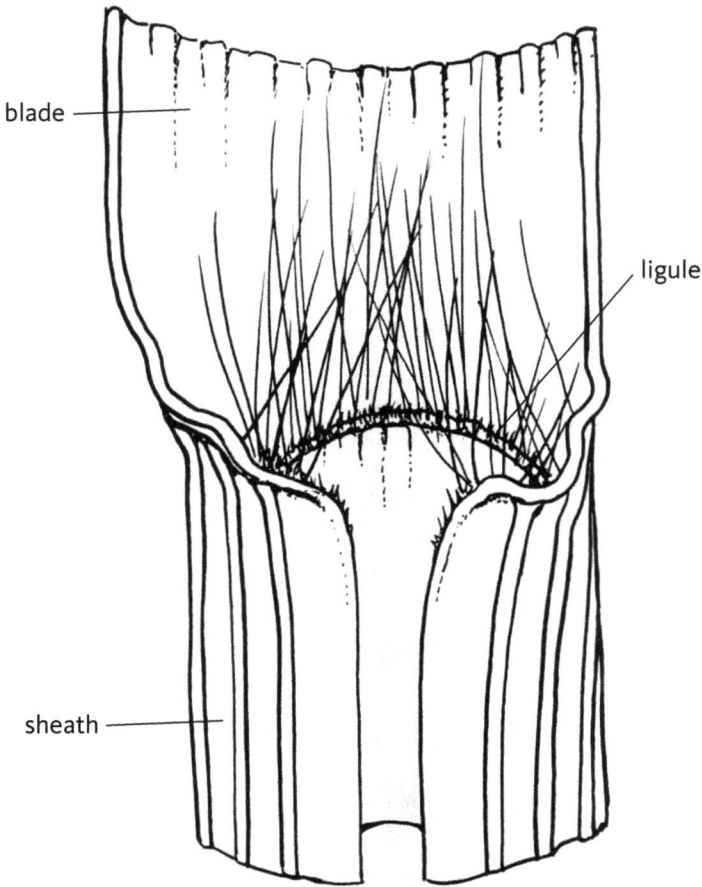

KEY CHARACTERS Large, erect bunchgrass, 1-2 m tall when mature. Blade narrow, long, rolled. Ligule hairy, small, 1/4-1/2 mm, long hairs at back 1-3 mm.

VERNATION Curled. **BLADES** Rolled, drooping, narrow, pointed; glabrous, veins each side of midrib 3-4; ribs indistinct, midrib not prominent; margin glabrous; width 1 mm, length 10-25 cm. **AURICLE** None. **LIGULE** Hairy, 1/4-1/2 mm, hairs on back of ligule 1-3 mm. **COLLAR** Margin pubescent, divided. **SHEATH** Round, glabrous, papery margin. **NODE** Glabrous. **INTERNODE** Glabrous. **roots** Short rhizomes. **CULM** Round, not branched.

HABITAT Dry, sandy to gravelly places, usually where alkaline.

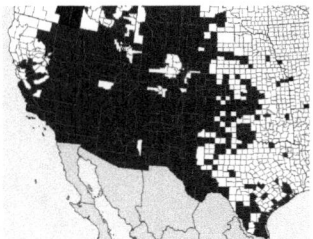

68 *Sporobolus compositus* (Poir.) Merr.
TALL DROPSEED

Sporobolus compositus (Poir.) Merr.
TALL DROPSEED

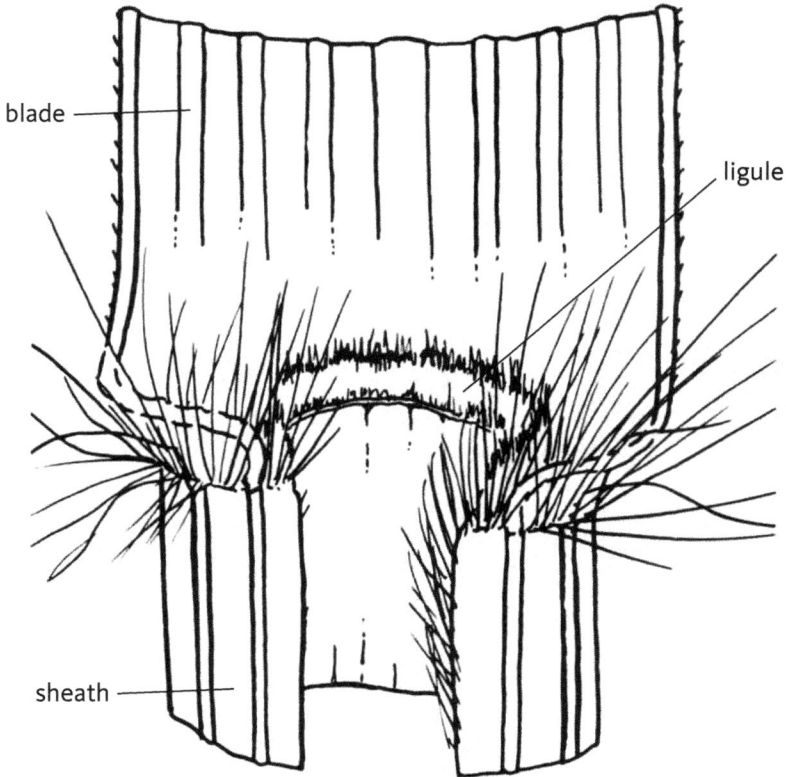

SYNONYM *Sporobolus asper* (Beauv.) Kunth

KEY CHARACTERS Erect, medium-sized bunchgrass. Blade flat, glabrous. Collar tufted. Sheath occasionally with marginal hairs. Panicles included well in sheath. Culm usually branched.

VERNATION Curled (sometimes reported as clasping). **BLADES** Usually flat, narrow, pointed; glabrous, rough ventrally, soft; veins each side of midrib 2–3; margin toothed; ribs prominent ventrally and dorsally; width 2–3 mm, length 5–20 cm. **AURICLE** None. **LIGULE** Hairy, short. **COLLAR** Tufted. **SHEATH** Margin hairy, 2–3 mm, usually pinkish at base. **NODE** Glabrous. **INTERNODE** Glabrous. **ROOTS** Fibrous. **CULM** Round, frequently branched at node.

HABITAT Roadsides and along railroads, beaches, cedar glades, pine woods, prairies, usually somewhat disturbed and where semi-open.

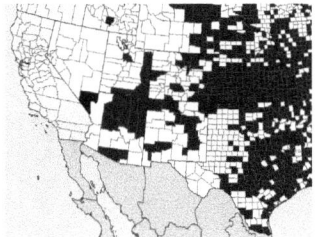

Sporobolus cryptandrus (Torr.) Gray
SAND DROPSEED

Sporobolus cryptandrus (Torr.) Gray
SAND DROPSEED

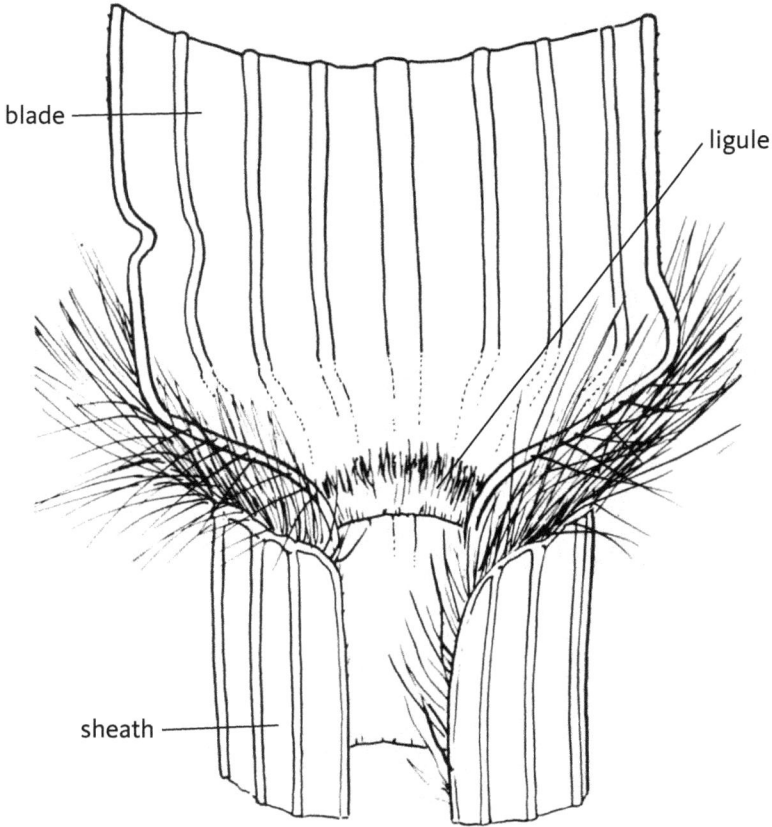

KEY CHARACTERS Erect, medium-sized bunchgrass. Blades flat, long. Panicle included in sheath. Ligule and collar bearded. Culms not branched.

VERNATION Curled. BLADES Flat, wide, long; glabrous; veins each side of midrib 3–4; ribs prominent; margin toothed; midrib not prominent; width 3–8 mm, length 15-20 cm. AURICLE None. LIGULE Hairy, hairs 2–3 mm on margin. COLLAR Conspicuous tufts of hair on margin, 2–3 mm. SHEATH Hairy often on one margin, the panicle more or less covered by the sheath. NODE Glabrous. INTERNODE Glabrous. ROOTS Fibrous. CULM Round, not branched.

HABITAT Sandy soils and washes, rocky slopes and calcareous ridges, roadsides, salt-desert scrub, pinyon-juniper woodlands, yellow pine forests, and desert grasslands.

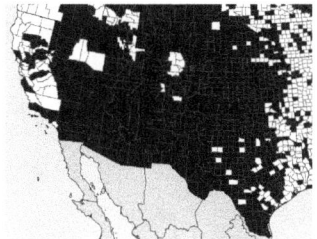

Sporobolus giganteus Nash

GIANT DROPSEED

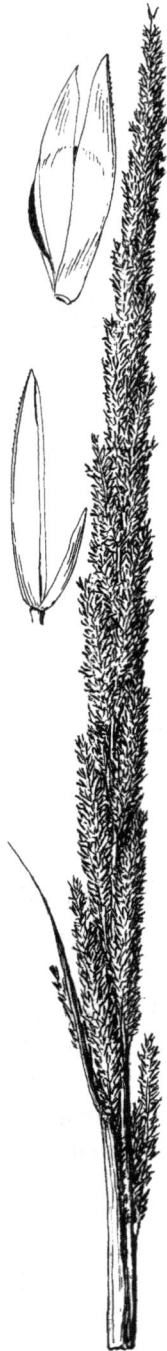

Sporobolus giganteus Nash
GIANT DROPSEED

blade

ligule

sheath

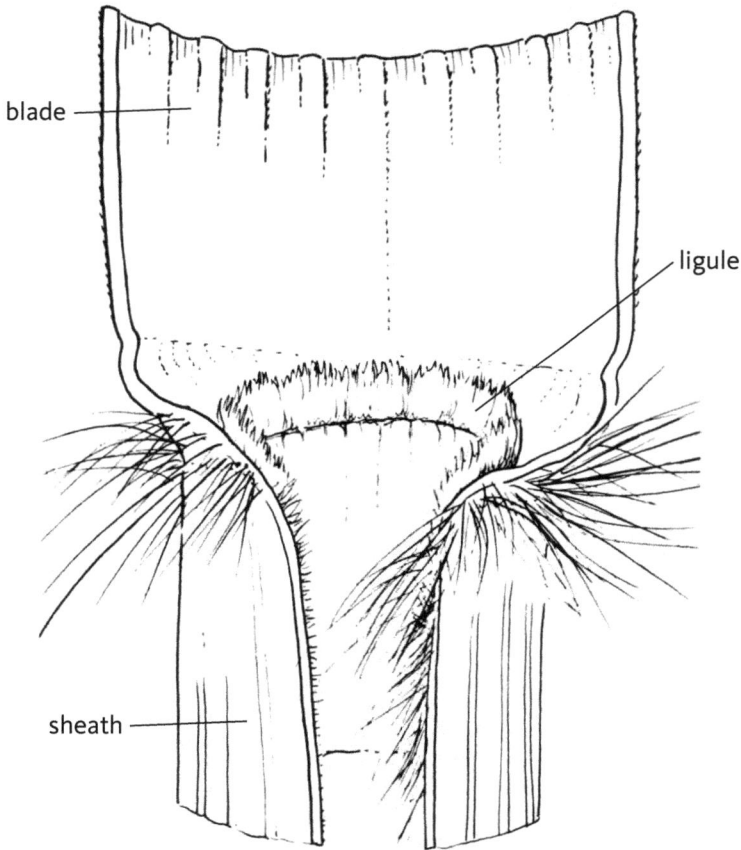

KEY CHARACTERS Large, erect bunchgrass. Ligule of hairs to 1 mm. Blade long, usually rolled.

VERNATION Curled. **BLADES** Flat to rolled, erect; rough ventrally, stiff; veins each side of midrib 3-4; ribs prominent dorsally; margin toothed; midrib not prominent; width 2-8 mm, length 15-35 cm. **AURICLE** None. **LIGULE** Hairy, short, 1/2-1 mm. **COLLAR** Glabrous, margins tufted, distinct. **SHEATH** Round, one margin usually hairy, veined. **NODE** Glabrous. **INTERNODE** Glabrous. **ROOTS** Fibrous. **CULM** Round, not branched.

HABITAT Sand dunes, sandy areas along rivers and roads.

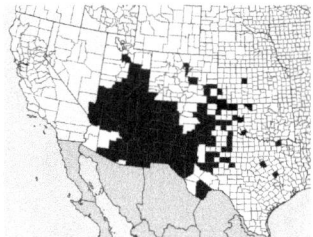

Sporobolus wrightii Munro ex Scribn.
SACATON

Sporobolus wrightii Munro ex Scribn.
SACATON

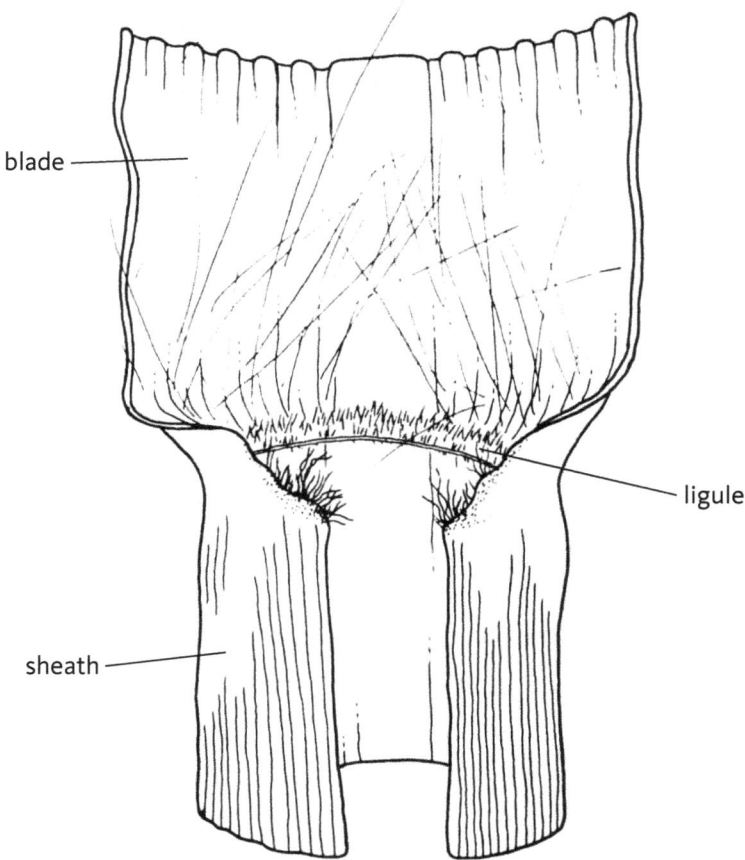

blade — blade

ligule — ligule

sheath — sheath

KEY CHARACTERS Large, erect bunchgrass. Blades flat, long, wide. Ligule short, hairy.

VERNATION Curled. **BLADES** Flat, drooping, long, narrow, pointed; glabrous; veins each side of midrib 3; ribs prominent ventrally; margin toothed; midrib prominent ventrally; width 3-6 mm, length 45-60 cm. **AURICLE** None. **LIGULE** Hairy, small, 1/2 mm, obtuse-lacerate. **COLLAR** Glabrous. **SHEATH** Glabrous, veined, round. node Glabrous. **INTERNODE** Glabrous. **ROOTS** Fibrous. **CULM** Round.

HABITAT Moist clay flats and on rocky slopes near saline habitats.

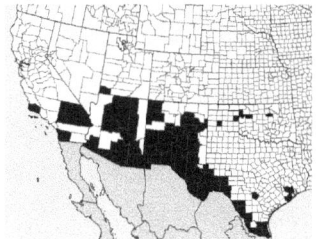

71

72 *Tridens muticus* (Torr.) Nash
SLIM TRIDENS

Tridens muticus (Torr.) Nash
SLIM TRIDENS

blade

ligule

sheath

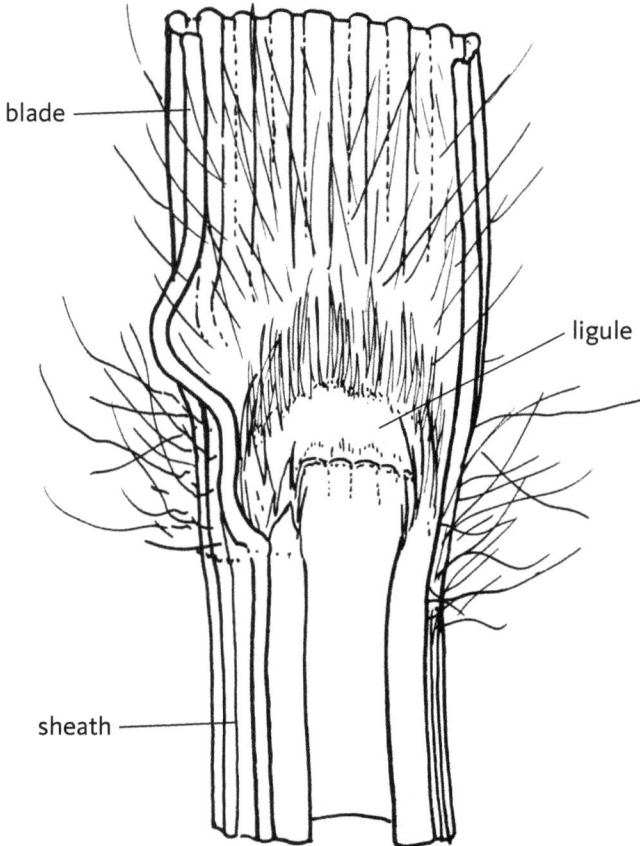

SYNONYM *Tridentopsis mutica* (Torr.) P.M. Peterson

KEY CHARACTERS Erect, small bunchgrass. Blade usually rolled. Ligule hairy and small. Node pubescent.

VERNATION Curled. **BLADES** Rolled, erect, narrow, pointed; with occasional glandular hairs dorsally; scabrous; midrib prominent dorsally; veins each side midvein 4–5; width 2–4 mm, length 8–20 cm. **AURICLE** None. **LIGULE** Hairy, small, 1/2–1 mm. **COLLAR** Hairy. **SHEATH** Open, hairy on margin, round. **NODE** Pubescent. **INTERNODE** Glabrous. **ROOTS** Fibrous. **CULM** Round.

HABITAT Dry, sandy or clay soils.

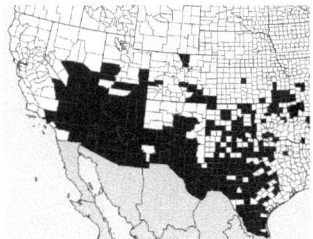

Zuloagaea bulbosa (Kunth) Bess
BULB PANICUM

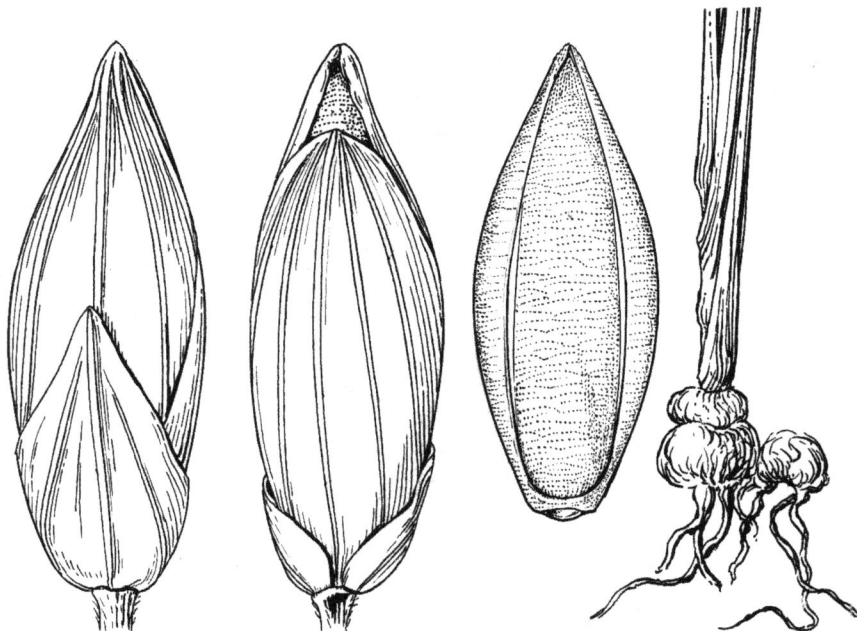

Zuloagaea bulbosa (Kunth) Bess

BULB PANICUM

blade

ligule

sheath

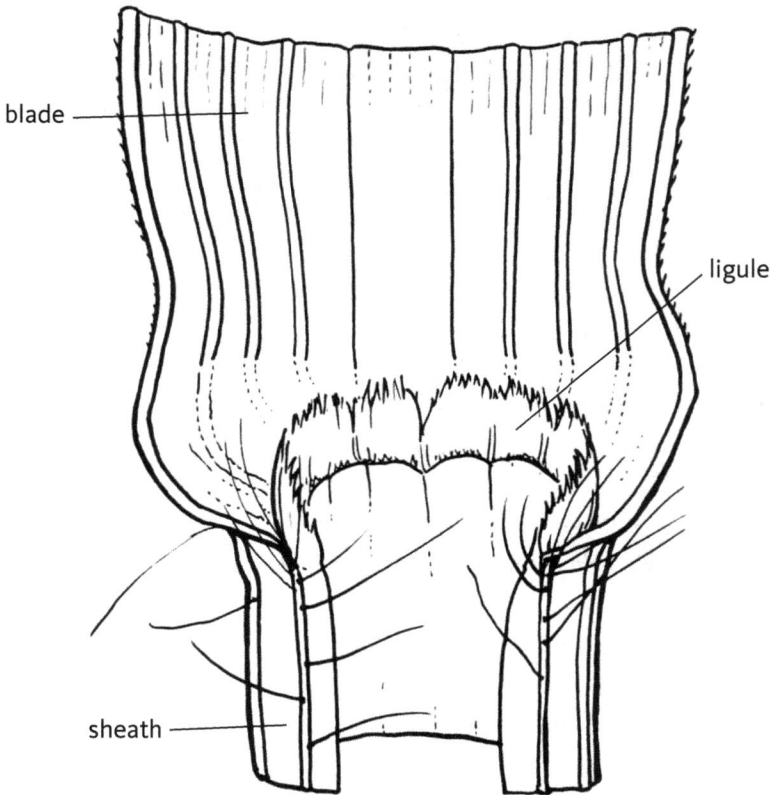

SYNONYM *Panicum bulbosum* Kunth

KEY CHARACTERS Small, erect bunchgrass. Bulbs prominent. Occasional glandular hairs on blade and sheath.

VERNATION Folded. **BLADES** Folded to flat, drooping, narrow, pointed; occasional glandular hairs near ligule; veins each side of midrib 3–4; rib prominent ventrally and dorsally, margin white, midrib prominent dorsally, width 2 mm, length 10–20 cm. **AURICLE** None. **LIGULE** Membranous, 1 mm, truncate-ciliate. **COLLAR** Glabrous, divided. **SHEATH** Elliptical, occasional glandular hairs near collar. **NODE** Glabrous. **INTERNODE** Glabrous. **ROOTS** Corm, rhizomes. **CULM** Flat, not branched.

HABITAT Roadside ditches, gravelly riverbanks, moist mountain slopes, often under ponderosa pine and oak.

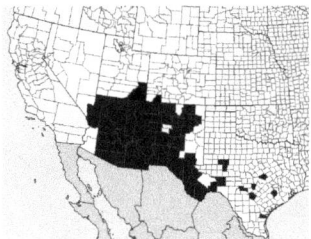

REFERENCE PUBLICATIONS

Anderton, Laurel K., and Barkworth, Mary E. (eds.) 2015. *Grasses of the Intermountain Region*. University Press of Colorado, Boulder.

Barkworth, Mary E., Anderton, Laurel K., Capels, Kathleen M., Long, Sandy, Piep, Michael B. (eds.) 2015. *Manual of Grasses for North America*. University Press of Colorado, Boulder.

Copple, R. F., and Aldous, A. E. 1932. *The Identification of Certain Native and Naturalized Grasses by Their Vegetative Characters*. Kansas Agric. Exp. Stn. Tech. Bull. 32.

Gould, F. W. 1951. *Grasses of the Southwestern United States*. University of Arizona Biol. Sci. Bull. 7, 352 p., illus.

Harrington, H. D., and Durrell, L. W. 1944. *Key to some Colorado Grasses in Vegetative Condition*. Colorado Agric. Exp. Stn. Tech. Bull. 33.

Hitchcock, A. S. 1950. *Manual of the Grasses of the United States*. U. S. Dept. Agric. Misc. Publ. 200, Ed. 2, revised by Agnes Chase.

Humphrey, Robert R. 1958. *Arizona Range Grasses*. Ariz. Agric. Exp. Stn. Bull. 298.

Kearney, T. H., and Peebles, Robert H. 1951. *Arizona Flora*. University of California Press, Berkeley.

INDEX

Synonyms listed in italics.

INDEX

Synonyms listed in italics.

INDEX

Synonyms listed in italics.

www.ingramcontent.com/pod-product-compliance
Lightning Source LLC
Chambersburg PA
CBHW051730020426
42333CB00014B/1249